Leftovers is dedicated to my darling daughter Romelia, with whom I was pregnant when I wrote it, and who slept next to me as I edited it over many late nights. The thought of your future has made the need to build a less wasteful food system strikingly more apparent, and I hope this book contributes, at least in its small way, to helping this happen.

LEFTOVERS

ELEANOR BARNETT is a historian with a PhD from the University of Cambridge, who currently holds a Leverhulme Early Career Fellowship at Cardiff University. Her work uses food as a lens through which to access the daily lives of ordinary people as well as wider cultural, economic, political and religious historical processes. As @historyeats on Instagram, she posts daily food history stories, paintings and objects from across the world to a wide audience and she is a regular contributor to other public-facing media outlets. *Leftovers* is her first non-fiction title.

ELEANOR BARNETT

LEFTOVERS

A HISTORY OF FOOD WASTE & PRESERVATION

HEAD ZEUS

An Apollo Book

First published in the UK in 2024 by Head of Zeus,
part of Bloomsbury Publishing Plc

9 7 5 3 1 2 4 6 8

A catalogue record for this book is available from the British Library.

ISBN (HB): 9781803281575
ISBN (E): 9781803281551

Typeset by Ed Pickford

Printed and bound in Great Britain by
CPI Group (UK) Ltd, Croydon CR0 4YY

Head of Zeus Ltd
First Floor East
5–8 Hardwick Street
London EC1R 4RG

WWW.HEADOFZEUS.COM

CONTENTS

INTRODUCTION

Lying in bed, the new Lord Mayor of London, John Key, is attacked by a horde of furious animals. A giant frog, loincloth covering its modesty, wields a sharp spear which has already been used to pierce the pink bodies of three smaller frogs as if on a spit. Spurred on by a gang of squawking poultry, a lobster pinches the man's nose with its massive claw, as a hefty sea turtle compresses his chest. Eyes clenched in pain and panic, the Lord Mayor desperately fires two bottles of bubbly – 'Wrights Cham[pagne]' – into the throng of his nightmare, as if they were pistols.

This fantastical image (see Plate 1) comes from a satirical lithograph of 1830, created in response to the last-minute cancellation of the great banquet which had been planned to inaugurate John Key into his mayoral office. In the image, a list resting on the bedside table informs us of some of the delectable dishes that had been prepared for the feast: 'Roast beef & à la mode', 'Veal and Mutton', 'Pork and Venison', 'Pheasants & Pigeons', 'Lobster & Sturgeon', 'Turkey and Capon', 'Frogs à la Crapodine', 'Shrimps in Pots', to name but a few. Meanwhile, the huge turtle reminds viewers of turtle soup, then the most prestigious of dishes which required live green sea turtles, reportedly weighing up to 600–700 pounds each, to be imported into Britain all the way from the West Indies.[1] The artist imagined all these creatures, their lives needlessly squandered for a banquet that never happened, returning to haunt the Lord Mayor's dreams.

As the Lord Mayor's Nightmare tells us, wasting food has long been a morally charged issue. The living things, whether sentient or vegetative, that become our food are created by the laborious processes of cultivating, tending, nursing, fattening, pruning, feeding, watering, slaughtering or harvesting. In the journey from farm to fork, they consume a portion of the world's finite resources. It takes 50 litres of water to produce one orange, for example.[2] And while a kilo of wheat requires 500–4,000 litres, a kilo of meat gobbles up 5,000–100,000 litres of water.[3] Today, fertilisers and pesticides are also spread generously on our food crops, which are grown on land requisitioned from nature using gas-guzzling machinery. To feed the animals whose meat and dairy products we simply throw away in UK and US households, plus that expelled from retail and food services in the US, we use up 8.3 million hectares of agricultural land. Growing soybeans, which become meal to feed these animals, is a leading cause of deforestation, especially in the Amazon.[4] This is to say nothing of the time, labour and skill needed to gather and transform these raw ingredients into meals. Wasting food means wasting life, but also the huge amounts of energy and resources that go into making it.

What's more, wasting food is only a possibility if you have plenty of it. Most Londoners could only dream of the abundance of meats, pies, soups, tarts, salads, cakes, fruits and sweets that would have been prepared for the Lord Mayor's banquet of 1830, had it occurred. Likewise, as an ungrateful child forced to swallow down those last few forkfuls of peas, I, like doubtless many others, remember that textbook parental admonishment: 'there are children dying in poorer countries who would kill for this food'. Today, food wasted is sustenance that could otherwise go into the mouths of the 842 million people across the world afflicted with chronic hunger.[5] Wasting food means wasting the money that was spent buying it, money that would certainly be of use in other hands. In 1830, one subject angrily wrote an open letter to William IV, whose customary appearance at the Lord Mayor's banquet had been cancelled at the last minute, condemning the fact that the king did not 'once check the pompous propensity to profligacy in the expenditure of the citizens' money'.[6] In the present day, food worth $1 trillion a year is wasted globally, while in Britain each of our households will waste on average around £500 a year needlessly on entirely edible food,[*] costing us £13.8 billion on a national scale.[7]

Over the last half-century food waste has garnered increasing public attention as we've gradually come to terms with the extent and urgency of man-made climate change. Perhaps you are already aware of the shocking statistic: across the world, a third of all the food we produce is wasted.[†] Left to rot and decompose, if all the food waste in the world was a country, it would be the third largest emitter of greenhouse gases after China and the United States. Meanwhile, in its creation, all this wasted food guzzles up 25 per cent of the world's fresh water and 10 per cent of global energy consumption in total, while the area of land needed to grow it is the size of China.[8] In the UK, this means over 4.5 million tonnes of edible food is wasted at home every year alone, the same CO_2 equivalent as 4.6 million return flights from London to Perth.[9] Unsurprisingly, reducing food waste is an important part of the global fight to lessen

---

* 'Avoidable food waste', meaning food or drink that was edible at some point before it was thrown away.

† 14 per cent of all food produced globally is lost between harvest and retail; 11 per cent of food produced is then wasted in households, 5 per cent in the food service industry and 2 per cent in retail. It is important to note that these percentages are not perfectly comparable: the first figure is a percentage of all agricultural production, whereas the second set of figures does not include animal feed; the latter set also includes inedible parts whereas the former does not. United Nations Environment Programme, *Food Waste Index Report 2021* (Nairobi, 2021), pp. 70–1.

the devastating impacts of climate change. There is little time left for the United Nations to reach its target – set in 2015 – of reducing food waste by 50 per cent across the world by 2030.*

The reasons that we waste food are numerous and complex. Having spent too long sat at my desk drinking tea and eating chocolate last week, I was determined to be healthy when I did this week's food shop, buying lots of promising – yet perishable – vegetables and other fresh ingredients. Starting the week as I meant to go on, I cooked a delicious and nutritious stir-fry for my family on Monday night, but after overestimating the quantity out of a fear of disappointing hungry stomachs, I ended up with leftovers. On Tuesday these scraps suddenly didn't look so appealing, so I cooked a cheesy pasta dish instead (no leftovers there!). On Wednesday, running out of time to make my lunch at home, I grabbed a quick bite out between meetings. Meanwhile strawberries left in their punnet were starting to sport white fluffy cloaks. By Thursday I'd thrown out the remains of Monday night's meal and the huge bag of kale I'd enthusiastically purchased was turning into an unappealing brown mush. On Friday, after a long work week, I felt too tired to cook and ordered a takeaway, forgetting about the salmon in the back of the fridge that needed to be used up by the end of the day. Aspirational shopping habits (buying what we think we *should* eat, not what we want or will eat), busy schedules that make us too tired to think about food management or that change at the last minute, as well as the difficulties of cooking for a small number of people, are all common culprits of food waste in modern life. Most food waste in the UK happens at home, but as we shall see over the course of this book, there are plenty of opportunities for food to be wasted before it even

---

* As set in the United Nations' Sustainable Development Goal 12.3 of 2015: 'By 2030, halve per capita global food waste at the retail and consumer levels and reduce food losses along production and supply chains, including post-harvest losses.'

gets to our plates: at farms, in transit and at grocery shops and restaurants.*

By contrast, we tend to look back at the past with rose-tinted glasses, as an unspecific time when people lived in harmony with the Earth, happily working the land to nurture their own food from the fields to the table. Perhaps we imagine a house-wife, smiling away in the kitchen to make wholesome meals for a large family, a family rosy-cheeked but not overweight, perfectly nourished by the natural resources around them. Until the last century, there were certainly no 'unnatural' pesticides poison-ing wildlife as we pollute habitats, no foreign ingredients flown across the globe in aeroplanes that leave gloomy smoke trails behind them, and no plastic packaging to clog up the Earth's river-arteries before it seeps into our contaminated seas. Yet this, of course, as the Lord Mayor's Nightmare suggests, is far from the whole truth. In exploring the past, we must endeavour to under-stand our ancestors' ways of life on their own terms, even as we draw comparisons and lessons – in both what to do and what *not* to do – from them. So why was food wasted in the past? Why and how did people try to prevent food waste? What did they do with the food waste that they did create? And what does all this tell us about their wider lives and values? This is a book about

---

* According to one recent estimate for the UK, consumption (in household or hospitality) accounts for 46 per cent of total waste, primary production (agriculture, farming and handling and storage at the post-harvest stage) for 28 per cent, food processing and manufacture for 17 per cent and food distribution (in wholesalers and retailers and transport) for 9 per cent. Harish K. Jeswani, Gonzalo Figueroa-Torres and Adisa Azapagic, 'The Extent of Food Waste Generation in the UK and its Environmental Impacts', *Sustainable Production and Consumption,* 26 (2021), p. 538. Excluding waste in primary production, WRAP estimates that household food waste accounts for 70 per cent of the UK total, manufacturing 16 per cent, hospitality and food service 12 per cent and retail 3 per cent. WRAP, 'Food Surplus and Waste in the UK – Key Facts', December 2022, https://wrap.org.uk/sites/default/files/2023-01/Food%20Surplus%20and%20Waste%20in%20the%20UK%20Key%20Facts%20December%202022.pdf.

the history of our complex and changing relationship to wasting food and kitchen thrift.

To begin, let's turn to the Great Nottingham Cheese Riot of 1766. On 2 October at the city's famous Goose Fair, violence broke out in response to the inflated price of cheese as local citizens tried to prevent the Lincolnshire merchants who had bought this precious commodity at the market from taking it away with them. Soldiers had to be employed to restore order as fierce Nottinghamians looted warehouses and shops, rolling gigantic cheese wheels across town (and knocking down another unfortunate mayor in the process). One farmer, William Egglestone, reportedly even died of a shot wound while gallantly guarding his cheese wheels.[10] Over a hundred different food riots took place in 1766, the third year in a string of failed harvests, which caused the price of grain – and therefore ordinary people's daily bread – to rise to uncomfortable heights. In the Wiltshire village of Bradley, for example, a mob descended upon the local mill, seizing the flour, corn, horse beans and lime housed there, and destroying everything else they found. The next day they marched on Beckington's mill but were met with a barrage of gunfire. Angered by rumours that the mill owner, Mr Carpenter, had exported precious stocks of grain at a time when prices were so high, the mob rallied fifteen guns of their own from nearby villages. After an exchange of ammunition, the riotous crowd chose to set fire to the mill instead and the owners fled to protect their lives.

As grain became dearer, the cost of raising animals and animal products like meat and dairy rose in response. As well as stealing flour and grain, like the Nottingham rioters, the Wiltshire mob turned their attention to cheese, breaking into the homes of cheese wholesalers to seize their produce, believing that these traders were unjustly pushing up prices by buying up local stocks.[11] At the most basic level, the tumultuous events of 1766 act as a reminder that the provisions of the Earth are not guaranteed to us, even though today supermarket shelves seem overwhelmed with an endless bounty of jars, pots, packets and boxes. The truth is that in unfavourable weather conditions, crops will fail to sprout from the soil, or droop and wither away back into the ground, with disastrous social, economic and political consequences.

What's more, the unending fight to protect our food supplies from foraging animals, and creepy-crawlies like bugs, worms and slugs, has been a constant in human history. Even the seemingly holiest of foods – the bread of the Eucharist, which Catholics believe contains the body of Christ – could fall prey to the curious nibbles of a church mouse, an unnerving fact that added fuel to England's Protestant Reformation in the sixteenth and seven-teenth centuries.[12] Rooks, crows and jackdaws were described around the start of the 1700s by one agricultural writer, Leonard Meager, as the 'great Devourers of Grain'. In his book, *The Mystery of Husbandry*, Meager advised the frustrated farmer to shoot some of the pests, 'hang them on Poles' across the field and cover the bounds with gunpowder, blowing some of it up and hoping the smell would dissuade the rest of the birds.[13] 'Bash!' said a rat-catcher centuries later. 'One more of Hitler's helpers was removed.' In the Second World War, 'protect[ing] our country's grain stacks' by catching and killing rats was an important job carried out by women on the home front to keep the nation supplied with bread at a time when food supplies were under threat.[14]

From the moment food is harvested or slaughtered, it risks becoming inedible as it begins to ferment, rot and decompose. This in turn attracts further interested parties from the animal kingdom. According to a poster published by the British Museum in 1918, the fly is 'one of man's greatest enemies', which feasts on decaying food and then vomits it up onto the food that we eat, depositing 'filth and germs'. By this date, physicians had begun to recognise the fly as a carrier of disease. Scientists were also indebted to Louis Pasteur, a French chemist who proved in the mid-nineteenth century that spoilage was caused by airborne microorganisms. His method of increasing the temperature of a liquid to destroy pathogens – which is essential to how we keep our milk fresh today – was named 'pasteurisation' after its inventor. Before Pasteur's work, maggots were thought to appear spontaneously in rotting meat when it was exposed to the air. Relatively recent scientific advances, therefore, have revolutionised our understanding of how our food decays and wastes away. Though our ancestors were no less aware of the need to preserve and protect their food from natural forces, in a world where

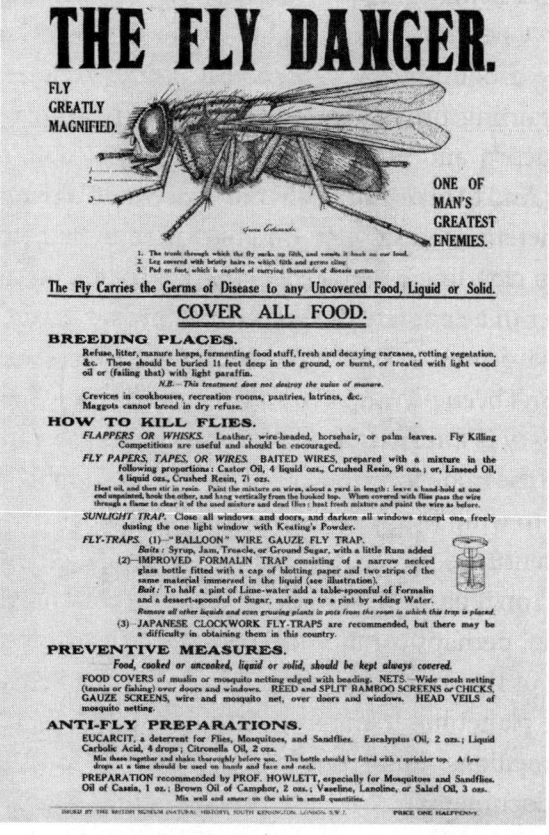

Figure 1. 'The Fly Danger': a poster published by the British Library and illustrated by Grace Edwards (1918), describing how flies are attracted to food waste and how to prevent them, for example by covering food with 'muslin or mosquito netting edged with beading'.

microscopic life-forms were largely elusive, the idea of food waste and its relationship with human health was understood within a scientific and medical framework entirely different to our own.

To ward off the attacks of organisms big and small, the food we eat today is kept fresher for longer using a whole host of modern technologies. At the farm, fruit and vegetables are protected from disease and demise with artificial fertilisers, herbicides, insecticides and fungicides. To slow spoilage and maintain its fresh appearance, nitrates, sulphites and organic lipophilic acids like sorbates and benzoates are commonly added to our food as it is processed. Chemical compounds with intimidating names like tert-butylhydroquinone and butylated hydroxytoluene are found in fat-containing processed foods, such as margarines and oils, crisps, popcorn and doughnuts, on account of their antioxidant qualities.[15] And to take one further tongue-twisting example, ethylenediaminetetraacetic acid is used in cans of drinks to prevent the liquid from clouding when it is exposed to the metal container.[16]

Whether in a can, tub, or box, go to the supermarket today and you'll be hard-pressed to find food that, once harvested and processed, hasn't been packaged in some way. The use of plastics in the food industry is a relatively new phenomenon, booming in popularity from the mid-twentieth century when it was still common for the grocer to wrap food in paper, and fill up customers' own jars or jugs. For centuries, the housekeeper visiting a food market would have relied on a basket to store the fish, vegetables and spices that they bought, perhaps using a linen cloth to wrap fresh meat in. Our awareness of the damaging effects that modern plastic packaging has on the planet has been bolstered in recent years by numerous public campaigns. From David Attenborough's stunning *Blue Planet II* documentary of 2017, for example, I still remember the unnerving image of a magnificent albatross feeding its young plastic rice packets, and the distressing fact that micro-plastics have become so concentrated in the natural food chain that dolphins' breast milk has in some instances become fatal to their calves.[17]

From New Year's Day 2022, the French government banned plastic packaging for most fruit and vegetables as part of its wider commitment to phase out single-use plastics by 2040.[18] In Britain, shops that require the consumer to fill up their own reusable containers are popping up once again. Still, the grocery sector is responsible for over half of the 1.5 million tonnes of consumer plastic packaging that is used in the UK retail sector each year, while UK supermarkets recycle only 30–34 per cent of the consumer plastic they use.[19] Even as we seek alternatives to plastic packaging, the vast majority of us still rely on packaged food, which works to protect our food from pests and prolong its fall into inevitable decay, as well as providing a handy surface on which to include legally required nutritional facts, expiry dates and brand information.

When we return home with our packaged food the first thing we do is put it in the fridge or freezer. By cooling down the food, refrigeration slows the growth of the pathogens that live on its surface. Early humans, who walked the chilly parts of the Earth with huge woolly mammoths, would surely not have failed to note the preserving power of the cold, even if they were ignorant as to how it worked. Cold, indeed, is such an effective preservative that an entire prehistoric woolly mammoth – including its erect 3-foot long penis (albeit squashed!) – was preserved in the ice of the Siberian permafrost when it was discovered in 1901.[20] Frozen mammoths have since emerged from ice so complete that several eccentric scientists and explorers of the last century were rumoured to have eaten them. (Not to worry: DNA analysis of the leftovers from the infamous 'mammoth banquet' at the Explorers Club in 1951 confirmed that the dish was in fact, mercifully, green sea turtle.[21]) Ancient cultures, such as the Chinese, Greeks and Romans, collected ice to cool drinks, a practice emulated by the wealthiest and most fashionable seventeenth-century Britons, who built ice houses with underground chambers lined with sawdust or straw as insulation. Yet home refrigerators first appeared hundreds of years later in 1913, and it was only in the second half of the last century that ordinary families in Britain could expect to make use of this technology in their own homes. How best to create artificial refrigeration was a conundrum that attracted many of the great and curious in history, from the Scottish professor William Cullen who published the first systematic study of refrigeration in the late eighteenth century, the American founding father Benjamin Franklin and his work on evaporation and cooling, to Albert Einstein, who patented a refrigerator design in 1930.

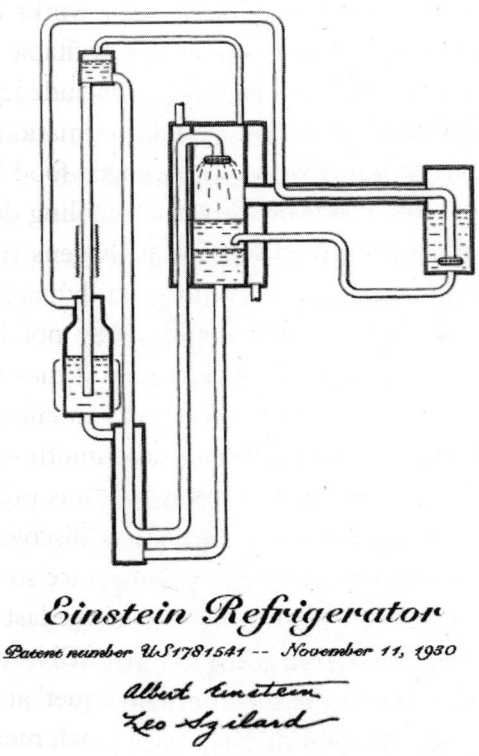

Einstein Refrigerator
Patent number US1781541 -- November 11, 1930

Albert Einstein
Leo Szilard

Figure 2. Albert Einstein's and Leo Szilard's US refrigerator patent, dated 11 November 1930. Though it was never a commercial success, the genius of the Einstein-Szilard machine was that it had no moving parts, removing the risk of leaking toxic gases, like ammonia, which were used as coolants. It also had no motor and used no electricity. In essence, the refrigerator worked by using an electromagnetic field to compress the refrigerant gas and move it around the fridge to absorb heat.

As we shall see over the course of this book, the need to preserve food before the advent of artificial refrigeration led to a myriad of delicious (and sometimes disgusting!) culinary inventions. Look in your kitchen cupboards and you'll be sure to find some preserved food items: tinned beans, tuna in brine, sun-dried tomatoes, jam, chutney, pickled vegetables, dried fruit such as prunes, raisins or currants. Even everyday staples like pasta and rice, tea and instant coffee are dried so that they can be stored for longer. Many of the basic techniques used to create these common ingredients date back centuries, if not millennia. In fact, the discovery of food preservation changed the course of human history, allowing our ancestors to give up a nomadic lifestyle in which they moved about continually in search of fresh food. Around 13,000 years ago, in the Middle Eastern region known as the Fertile Crescent, humans made their first permanent settlements. By domesticating crops and storing any surplus for future use, the food supply could sustain a larger portion of the community than was possible through hunting and foraging alone. With time, some well-fed settlers became priests, others became artisans and some specialised as fighters. And, eventually, with more time, came all the richness of society and culture.

As a species, our survival has depended upon preserving food during the glut of summer and autumn to feed ourselves through harsher winter conditions. Unsurprisingly, across the world examples abound of drying, salting, smoking, sugaring and fermenting to prevent food from going to waste. Indian ghee, used in a variety of dishes across the subcontinent, is a kind of preserved butter from which the liquids have been removed. Kimchi is a traditional Korean dish of salted and fermented vegetables that, in a process called Gimjang, was prepared in the autumn so that it could be enjoyed through the winter. In the high altitudes of the Andes in South America, even before the flourishing of the Incan Empire in the fifteenth century, indigenous peoples preserved potatoes by exposing them to freezing night-time temperatures, dehydrating them in the heat of the day, and stamping out the last drops of water. After several days and nights of this treatment, the hardened shrunken potato, called chuño, can last for months and years, and is rehydrated with boiling water when the time comes to consume it.* High in sugars which expel water, honey is traditionally used by the Kikuyu people in Kenya to preserve meats, which so treated become a delicacy called rukuri. In the far north, the Inuit people of northern Greenland make a special occasion food called kiviak (or kiviaq) by cramming many whole little auks (a type of seabird) inside the hollowed-out stomach of a seal. The birds are left to ferment over several months, and traditionally provide an important source of food over the course of the gruelling Arctic winter.

Many preserved foods continue to feature in our diets and even shape our concept of a national cuisine. Bacon, for example, takes centre stage in the traditional British fry-up. In medieval Britain, pigs were reared on a 'pannage' system in which they were left to graze on acorns, chestnuts or beech-mast in woods and forests. Though it would have tasted quite different to our modern version, bacon began as one of the ways to preserve the meat of pigs slaughtered in autumn to last through the cold season. A humorous anecdote recorded in the seventeenth century describes a certain Mr Hog being tried on account of 'a long career of crime' in the court of the Elizabethan judge Sir Nicholas Bacon. The prisoner pleaded for mercy on the grounds that the two men were related: 'Lord your Name is Bacon, and mine is Hog, and in all Ages Hog and Bacon have been so neer kindred.' Lord Bacon dismissed Mr Hog's pleas and sentenced the thief to death: 'for Hog is not Bacon

---

* I am indebted to Professor Rebecca Earle for teaching me about chuño when I was an undergraduate at the University of Warwick. Almost a decade later, I am still the proud owner of the chuño I made under her direction by – over the course of a week or so – repeatedly putting a potato in the freezer overnight before defrosting it during the day and squeezing out the moisture.

until it be well hanged', he quipped.²² This was an 'apophthegm', a saying that reveals an accepted truth, that the slaughtered carcass of a pig was hung to cool before it was cured into bacon by adding salt and hanging it out to dry once again.

For thousands of years, cheese has been produced as a way of preserving spring and summer milk supplies. Fermentation, the same microbial process that leads to the decay of food, creates cheese when the growth of specific bacteria is actively controlled and encouraged rather than being left to nature. Milk bacteria digest the milk sugar lactose and produce lactic acid, which, along with added rennet,* creates curds; these are then left to 'ripen' as the microflora gets to work, and the cheese is dried and salted. The Roman army famously marched on hard cheese, a forerunner to the hardy Parmigiano Reggiano, which can last for many years without spoiling.

Preserved foods, indeed, fuelled our ancestors as they travelled, explored and waged war, allowing the British to journey across seas to lands unknown to them which they might otherwise have been unable to feed from. For example, when the Scottish explorer Alexander Mackenzie made his celebrated journey across North America in 1793, becoming the first person recorded to cross the continent further north than Mexico, he was sustained in part on a diet of pemmican.²³ Mackenzie learned of this energy-rich food from indigenous people, who in parts of North America made pemmican (from the Cree Indian word for 'fat') by combining together strips of lean meat (most often bison, deer, elk, or moose), which had been dried in the sun and pounded between rocks into a powder-like consistency, with equal amounts of melted fat. Dried fruits like cranberries could also be added.

If many of the dishes that we eat today were created as ways of preserving food, others are the result of attempts to recycle

---

* Enzymes (especially chymosin) produced in the stomachs of ruminants. These days most commercial cheeses use chymosin from non-animal sources.

what we might consider food waste into new edible products. Thrifty cooks using up stale bread by soaking the crumbs in milk, perhaps adding in some sweetness, and baking the resultant mixture, is likely the origin of that school-dinner classic bread and butter pudding, for example.[24] Nineteenth-century cookbooks describe another traditionally British fare, cottage pie (this and the name shepherd's pie seem to have been used interchangeably at the time), as a good way of using up leftover meat. Once minced, the meat was covered in gravy or an onion sauce in a deep pie dish, and topped with mashed potatoes.[25] With its iconic lion logo, a tin of Lyle's Golden Syrup is a staple in many a kitchen cupboard. It was first created in the 1880s from experiments at a sugar refinery in Plaistow, England, undertaken by the company's chemist Charles Eastick, who was attempting to extract more crystalline sucrose from the waste treacle produced in the sugar refining process during a shortage of raw sugar.[26] And, love it or

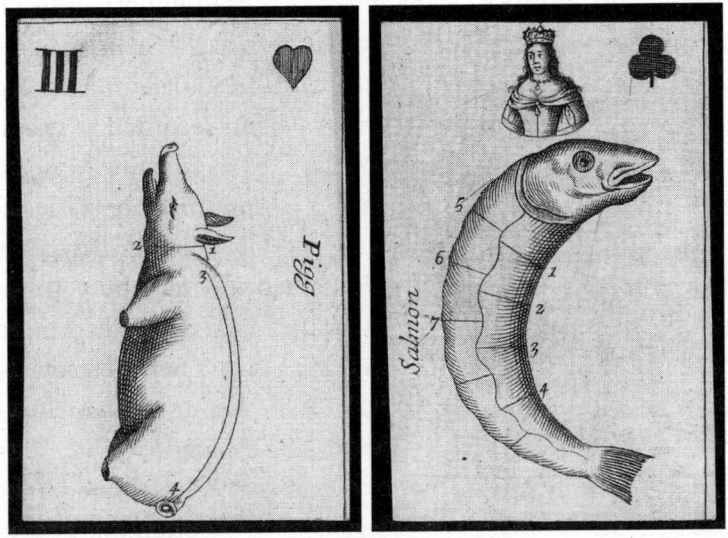

Figure 3. Examples from a pack of cards explaining how to carve meat in order to avoid 'an unthrifty wasting of it', from *The Genteel House-Keepers Pastime* (London, 1693).

hate it, Marmite was invented at the turn of the twentieth century at Burton-upon-Trent in Staffordshire as a means of using up the spent yeast from a nearby brewery.[27] Cooks have long instructed readers on how to make the most of the food they have, by using up all parts of an animal or reinventing leftovers in a new dish. From at least the seventeenth century, we have been reminded that 'wilful waste makes woeful want', a precursor to the now-familiar proverb 'waste not, want not'.[28]

So what about the food waste that we do throw away? It may seem out of sight out of mind, but the food that we discard from our homes and businesses has an afterlife. In fact, it is food waste – pips, shells, cores and (later) food packaging – that allows archaeologists to unearth the daily lives of past communities. Historian Tom Licence's excavation of late Victorian rubbish pits has uncovered the beginnings of a 'throwaway society', in which the reusable stoneware jars, pots and bottles that occasionally made their way into dumps after being chipped or smashed were replaced with mountains of disposable glass bottles and cans, marked with the logos of emergent brands.[29] Focusing on the more recent past, William Rathje's *Garbage Project*, which began in 1973, found perfectly preserved lettuce heads and a forty-year-old hot dog in the landfill sites of the United States, where a lack of light, moisture and oxygen often prevents food waste from breaking down. The researchers were surprised to discover an increase in beef waste during the US' beef shortage of 1973 to 1974 (from 3 per cent of all beef purchased in an average year to 9 per cent), suggesting that people concerned about meat supplies were stockpiling food, and – counterintuitively – then wasting it.[30] Think about the decisions that you make in your own life. Do you compost, separate your food waste into a caddy for collection, or send mouldy, rotting, smelly leftovers to be buried in landfill sites? Are you conscious of what happens to the packaging that your food comes in when you throw it away? This is a history, too, of these leftovers and their impact on people and planet.

Leftovers begins its story as Britain* was emerging from the medieval into the early modern period. In the following half a millennium, as we've travelled through eras of technological advance, industrial revolution, world wars, globalisation and pandemic, the way we eat has changed beyond recognition. If it seems like the relentless need to feed ourselves and our families consumes a great deal of time in the modern world, imagine how much greater the burden would be if you grew your own fruit and vegetables, milked your own cows, made your own butter and cheese, brewed your own ale or beer, and had no refrigerator in which to store it all. Despite these monumental differences in our lifestyles, the social, political and environmental impacts of food waste echo through the ages and warrant further exploration. Our discussion here includes what sociologists[31] might call 'avoidable food waste' that could have been eaten – from the farm, the manufacturer, right down the food supply to the retail and household level; 'potentially avoidable' waste, such as potato skins or bread crusts (whether or not to eat these items is often a matter of cultural norms or personal choice); as well as 'unavoidable waste' like bones, eggshells, tea bags and other inedible by-products.† This is a history of leftovers, of the food that we couldn't eat, that we wasted, or which was consumed by the unrelenting forces of nature itself. Yet, as this initial foray into the history of food waste hints, this is also a story about our

---

* As this book covers such a long period of time, I use Britain here to refer to England, Scotland and Wales. In reality, Britain was not formed until the 1707 Acts of Union, which brought together the kingdoms of England (including Wales) and Scotland. The United Kingdom included Ireland from 1801 and from 1922 referred to Britain and Northern Ireland.

† Sociologists also often distinguish between 'food loss', that which spoils or is otherwise wasted in farms and manufacturing processes; 'food waste', from food that was sold for human consumption and wasted at the retail or household level; and 'food surplus', when an excess of food is produced and is at risk of going to waste as a result. All of these categories are included in this book.

changing morals, of charity and community, poverty and wealth, the advancement of public health, human understandings of nature and our place in the natural world, of exploration, discovery, culinary ingenuity, invention and scientific advancements. If, as the old adage goes, 'you are what you eat', we – our values and culture – are equally defined by what we *don't* eat.

1

PICKLES AND PIETY

To preserue Cowcumbers all the yeere.

Take a gallon of faire water, & a pottle of veriuice, and
a pint of bay salt, and a handfull of green fennell or Dill:
boyle it a little, and when it is cold put it into a barrel,
and then put your Cowcumbers into that pickle, and
you shal keep them all the yeere.

Sir Hugh Plat, *Delightes for Ladies* (1602).

This recipe for pickled 'cowcumbers' – the early modern
spelling for 'cucumbers' – was one of many methods for
preventing food waste recommended to Tudor and early
Stuart housewives in the numerous recipe and advice
books of the era.

———•———

Writing in 1660, chef Robert May instructed his readers to
form stiff pastry into the likeness of a stag that had been
shot with an arrow, and to fill the pastry carcass with bright-
red claret wine to mimic its blood. Another dish should display
an intricate 'paste-board' (cardboard) and pastry battleship
complete with model cannons filled with real gunpowder, and
another still was to be forged into a castle 'with Battlements,
Percullices [portcullises], Gates, and Draw-Bridges', both ship

and castle to be surrounded by piles of salt to represent water. May imagined that 'some of the Ladies' at the feast would 'be perswaded to pluck the Arrow out the Stag', causing the wine to spill and drench the damask table cloths. Delightful chaos would then ensue: the ladies would throw eggs at each other, only to find that they had been emptied of their natural contents and filled with fragrant rosewater; a mock battle would commence between the pastry battleship and castle; and live birds and frogs that had been hidden in gilded pies would surprise the unsuspecting dinner guests, causing the 'Ladies to skip and shreek'. In this exuberant imagined culinary display, May was seeking to recreate the great Christmastide feasts of the Tudor and Jacobean eras, a time before, he said, 'good House-keeping had left England'.[1]

In this period, indeed, elite kitchens would have worked overtime for twelve full days between Christmas and Twelfth Night (5 January) to cater for the most indulgent season of feasting and partying. At the royal court, King Henry VIII was presented with far more food than he could possibly have eaten. The Christmas menu would have included a staggering array of meat dishes, including beef, mutton, veal, pork, chicken, goose and game birds like quail, woodcock and pheasant. Some meats were boiled, others roasted for hours over a spit and served with their own juices enriched with vinegar or wine, and others were minced and mixed with expensive currants sourced from the Levant and baked into pies. Alongside a range of garden vegetables, fruit tarts and jellies, these innumerable plates were served in buffet form alongside the centrepiece of the Christmas feast: a great gilded boar's head stuffed with its own meat (or brawn) which had been soaked for weeks in Continental wine and eastern spices (such as black pepper, cloves and mace), garnished with bay and rosemary, and served with a generous smearing of mustard. Swans and peacocks were also presented with great pomp at courtly feasts throughout the year, the roasted meats enveloped in the birds' feathers so as to take on a life-like appearance, and their beaks gilded (see Plate 3).

A noble feast like this would have been followed by a banquet, a term that then referred to a specific dessert course of sugar sculptures, biscuits, candied fruits and sweets (known as comfits) made from layers and layers of sugar syrup wrapped around small pungent seeds like caraway, anise, fennel and coriander. Much as May imagined, the Tudors and early Stuarts delighted in culinary burlesque: Henry VIII commissioned a sugar sculpture of a dungeon on a lake with swans; his leading statesman Thomas Wolsey created an entire sugar chessboard complete with moving pieces; and even the plates and cups at these events could be made with sugar and eaten once their contents had been consumed.[2] Sugar was made into a fragrant malleable paste with the addition of 'gum dragon' or tragacanth (the dried tasteless sap of *Astragalus* plants native to the Middle East), which was steeped in rosewater, and mixed with lemon juice and egg whites. 'Marchpane', the precursor to modern marzipan, made of sugary almond paste (again often flavoured with rosewater), was another popular choice. One recipe book of 1608 called for 'marchpane banqueting conceits' of pies, birds, baskets and walnuts in which a little piece of paper with a poem or saying could be hidden, an early modern version of a fortune cookie.[3] These treats would be decorated with a variety of natural food colourings: sandalwood for red and saffron for yellow, for example.

Finally, all this was to be washed down with copious amounts of alcohol, as William Harrison noted in his 1577 commentary on English diets: 'as all estates do exceed herein, I mean for strangeness and number of costly dishes, so these forget not to use the like excess in wine'.[4] As well as locally produced ale and beer, the rich imported a light-red claret wine from France, a sweet red sack from the Iberian peninsula, and muscadine from Italy, to which they were even inclined to add further spoonfuls of sugar. It is no surprise, then, that our culinary image of early modern England tends to be one of excess, gluttony and wastefulness, led by a greedy king with a bulging stomach.

At the other end of the social scale, years of poor harvests, especially in the 1590s, led more and more people into poverty and hunger. Pressures on food provision were compounded by the sixteenth-century boom in population, and the rise of enclosure, in which landowners fenced off common land so that labourers couldn't access it to pasture their livestock (sometimes even evicting them from their homes in the process). For the labouring poor this could mean the loss not only of their meat supply but also of income derived from their livestock, as in the wool trade. Such conditions led to unrest. At 9 p.m. on 21 November 1596, when England 'was in the grip of the worst subsistence crisis of the century', four poor men from Oxfordshire met on Enslow Hill hoping to ignite a rebellion and march on London.[5] With at least forty of their compatriots the men had visited the local lord lieutenant, Henry, Lord Norris, at Rycote, demanding that he help the poor as local grain reserves dwindled, and warning that if he did not act speedily, 'they would seek remedy themselves, and cast down hedges and ditches, and knock down gentlemen'.[6] Led by a local carpenter named Bartholomew Steer who was in the employment of Lord Norris at the time, the so-called Oxfordshire Rising aimed to do just that. Its ambitions were quickly thwarted, however, and its members arrested, tortured and killed (either in prison or executed for treason).[7] This attempted revolt came a year after around one thousand London apprentices had rioted on Tower Hill in response to rising food costs in the capital, where the price of flour had nearly tripled between 1593 and 1597.[8] The disastrous impact of this increase on the lives of ordinary people can only be imagined with the knowledge that bread was the most significant staple in early modern England, the true 'staff of life' on which every single meal was based. Poor harvests occurred with some regularity, falling every decade or so; earlier in the century the years 1527–9, 1549–51 and 1554–6 had been particularly challenging for England's inhabitants.[9]

Times of scarcity required almost everyone to adapt their normal diets to prevent food waste and to make the most of existing supplies. In his recipe book of 1596, *Sundrie New and Artificial Remedies Against Famine*, the eccentric inventor Sir Hugh Plat recommended 'Sweete and delicate cakes made without spice, or Sugar' to his middling and upper-class readers. Parsnip, dried and beaten into a powder, he said, would act as a sweetener instead of sugar and spices such as cinnamon, ginger, mace and nutmeg, which were among the most expensive ingredients of the era.[10] Cakes made of sweet root vegetables would be similarly recommended during the Second World War, another period of national food insecurity, when the British government's Ministry of Food published recipes for carrot-based Christmas pudding.[11] Compared to carrots, parsnips add a similar 'flavour and sweetness with a whole lot more fragrance', according to the *Great British Bake Off* winner Nadiya Hussain (whose parsnip and orange spice cake is a far-from-austere delight!).[12] Back in the Tudor kitchen, Plat further advised his readers to get through the time of scarcity by using cheap lentils and beans to make bread instead of the wheat loaves sought after by the elites, and by using up 'dry, hard, or stale grated bread' in a 'hochpot' (perhaps you might say 'hodgepodge' today), a kind of anything-goes soup made with water or a simple broth.[13]

These thrifty recipes are a long way from the wasteful rosewater egg-bombs and wine-soaked tablecloths described by Robert May. Yet, as we shall see, even during times of more bountiful harvests early modern society as a rule sought to avoid needless food waste. Labourers on farms relied on a simple diet of mixed-grain breads (wheat, rye, barley, oats, for example) and pottage, a thick soup filled with seasonal vegetables and leftovers from previous meals. It is notable, as the food historian Ken Albala writes, that on the relatively rare occasions that poorer people had fresh meat to eat, 'they would typically consume every last morsel in one form or another down to every organ and bone and even the blood'.[14] Most Tudor households, the vast majority of which were situated in rural locations, would have kept at least one pig. When the time came to slaughter it – an exciting and ritual event in most communities – the cottager would hold the creature down, and, plunging a knife into its side, collect the flowing blood in a bowl so that it could be made into blood pudding. Once the bristles had been removed and the carcass hung up, the animal was butchered. Hams and bacon, and the more heavily brined salt-pork, were cured for later consumption, while the pig's innards were fried and eaten straight away; the head, feet, ears and chitterlings (the large intestines) were commonly pickled; the head meat could be made into brawn; bones could be saved in brine to add flavour to sauces; while the rendered fat made a luxurious cooking medium.

Offal products were not only eaten by the poor in the early modern period. Richer people, too, enjoyed a much greater range of meat products than we commonly do today in the Western world (see Plate 4). Cows' tongues and calves' feet, for instance, were a regular feature at elite tables, even when they had a lot of other meat to eat. Lady Penelope, the widow of William, second Baron Spencer (part of the noble family who centuries later through Princess Diana would marry into British royalty), kept the accounts at the Althorp Estate in Northamptonshire after her husband died in 1636. On an ordinary week in July 1639 she recorded expenses for meat that included mutton, veal, lambs, bacon, pigeons, chickens, pullets (young hens), capons (castrated domestic cocks) and rabbits, as well as a 'paire of udders', marrow bones (those that contained the largest amount of marrow), two 'neates tongues' (ox tongues), five sheep heads and calves' heads.[15]

All these meaty products could be cooked into sumptuous dishes. According to one recipe book of 1584, calves' feet were made into a pie by first boiling and blanching 'the haire of[f] them', before being seasoned with luxurious spices including cloves and mace, pepper and sugar, along with prunes, currants, and dates.[16] Even the stock produced in the process of boiling the feet would not be wasted, but strained off to make calves' foot jelly, a delicacy created by combining the stock, egg whites and various spices.[17] It could be coloured, moulded and served on the banqueting table as a stand-alone dish. Meanwhile, sheep guts became the casings for sausages, and the animal's stomach was used to make the bags for boiled meat puddings, a technique that survives in more traditional Scottish haggis recipes today.

'The Accomplish'd Lady's Delight in Preserving'

Rich or poor, all people living in the early modern period were attuned to the seasonal booms and busts of the Earth's provisions. Whereas today we use artificial light to trick hens into laying eggs all year, right up into the twentieth century hens hardly laid in the coldest winter months. Milk, too, is today supplied all year by artificial means, given that cows need to have given birth recently to produce milk (just like us humans). In the Tudor and Jacobean eras, modern methods of artificial insemination had not yet been developed, so when cows were between calves they went through a dry period which could last from September until the spring when they might give birth again.[18] In terms of meat, veal was best in early spring, pork in late spring, and mutton in June. Martinmas, which fell in November (St Martin's Day is 11 November), was the traditional time to slaughter livestock and preserve it for consumption during the colder months. Preservation – of fruit and vegetables as well as meats – was key in maintaining a varied food supply in all four seasons and in preventing the waste and spoilage of seasonal produce.

Today, most of us preserve our food every day without even thinking about it, by storing perishables in the fridge and freezer, but artificial refrigeration was only just beginning to be explored in the sixteenth and seventeenth centuries. Early experimentation had found that a vessel of liquid could be cooled by moving it in a mixture of saltpetre* and snow or ice, the salt acting to reduce the temperature of such to below freezing.[19] But research into creating artificial cold was not without its risks. Sir Francis Bacon, famed for founding the scientific method of empirical investigation, died from pneumonia in 1626 after having spent too long trying to preserve meat by stuffing a fowl with snow![20] Not long afterwards, a group of male scholars working on scientific experiments to explain natural phenomena (including the functioning of hot and cold) founded the Royal Society for the Improvement of Natural Knowledge in 1660,[21] and many famous early scientists, such as Robert Boyle and Isaac Newton, would subsequently join its ranks.

Outside of this professional male realm, daily experiments in food preservation had long been conducted by women in the domestic setting. Well-to-do housewives in country estates wrote, collected and shared recipes in manuscript books which were passed down to female family members, neighbours and servants. Published cookbooks, with titles like *A Closet for Ladies and Gentlewomen*, or (later) *The Accomplish'd Lady's Delight in Preserving*, emerged for the first time in the mid-sixteenth century and, although initially written exclusively by men, they were marketed at elite women for use in the still-room.[22] This was a special room separate from the kitchen, fitted with slatted storage cupboards that had space for a charcoal heater for preserving and distilling foods and drinks. Books offering a range of advice on 'housewifery' also boomed in this period, including how to cultivate, cook and preserve foods, alongside instructions on cleaning and making medicines. While wealthy women could delegate much kitchen labour to servants and cooks, most

---

* A type of salt called potassium nitrate, which was found to be a useful preservative in the seventeenth century. Today, saltpetre more often refers to Peru saltpetre, or sodium nitrate.

worked tirelessly whisking, pounding, chopping and straining ingredients, all physical tasks from which we are today largely removed given our reliance on electric gadgetry.

Historians are increasingly recognising the role of women as generators of knowledge as they carried out these domestic tasks.[23] According to historian of science Simon Werrett, women's domestic experiments in 'thrifty' cookery and the scholarly male experiments of the Royal Society were really 'part of the same enterprise'.[24] One such inventive lady was Elinor Fettiplace, who came from an ancient gentry family, the Pooles of Sapperton in Gloucestershire. In 1589 she married Sir Richard, with whom she would eventually have five children. Lady Fettiplace ran their household at Appleton Manor near Oxford, and compiled a book of recipes that she dated 1604. Perhaps most numerous among her notes were instructions for making marmalades and fruit preserves, a task deemed most fitting for well-to-do ladies (and the middling classes who followed their example) as they were to be served on the banquet tables that closed luxurious dinner parties. Quince, a deliciously fragrant fruit that has somewhat since fallen out of favour in the UK, needs to be cooked and softened before being eaten and so was routinely made into marmalade. The word 'marmalade', in fact, comes from the Portuguese word for quince: *marmelo*. Around early October, Lady Fettiplace would gather the last of the quinces from the orchard at Appleton Manor, pare and core them and put them into a boiling syrup made from equal weights of quince, sugar and water: 'when the syrup comes to bee iellie, then they are doone', she wrote.[25] Though it could be served in pies and tarts, unlike today marmalade was also often eaten on its own. When it was boiled with sugar beyond its boiling point, the fruit mixture stiffened and became a thicker paste, which could be sliced into squares, and even stamped and shaped with elaborate decorative moulds, perhaps a floral pattern, the imprint of a coat of arms, or the image of a mighty stag.[26] Cinnamon, musk and rosewater were all familiar additional ingredients that appealed to refined seventeenth-century taste buds.[27] Marmalade made in this way was then stored in 'marmalade boxes', which could be kept for around twelve months and were often given as celebratory gifts at the turn of the New Year.

'To make oringe marmalad', according to Lady Fettiplace, you should first squeeze out the juice from the oranges and boil the rinds until they become soft. Next, 'bray', or grind them 'very smale in a morter', before returning the rinds to a mixture of sugar, water and the squeezed orange juice.[28] Thomas Dawson, author of *The Goodhuswifes Iewell* of 1587, recommended a different technique 'to preserue Orenges' that kept them more intact. First, the housewife should cut the oranges in half and 'let them lye in water foure or five dayes', changing the water once or twice a day. Next, the orange halves should be added to a boiling mixture of a quart (equivalent to between 0.95 and 1.16 litres) of water with sugar and a little rosewater and cinnamon, set over a fire. This should be done five or six times as the fruit absorbs the syrup.[29] Either cut up or whole with their peel intact, oranges preserved in syrup could then be stored in gallipots* or glasses. At this time, most oranges were of the bitter Seville variety, those that we would use today to make traditional English marmalade, rather than the sweeter Asian or 'China oranges' that had only arrived in Europe in 1529 and were not common until much later.[30] Oranges were a highly prized fruit in the early modern period, and by the end of the seventeenth century it had become fashionable to grow them in England in special orangeries. The great diarist Samuel Pepys, writing in 1668, described the orange and lemon trees of Lord Wotton's garden as 'the most noble that ever I saw'.[31] It is no surprise that housewives lucky enough to have access to oranges worked to ensure year-round bursts of their striking flavour.

Other fruits, including apricots, plums, peaches, apples, raspberries, gooseberries and cherries, were preserved in similar

---

* Small ceramic jars.

ways. This task kept ladies like Elinor Fettiplace and her maids occupied throughout the year, with imported oranges in season in January and February, green plums and gooseberries in early summer, and peaches and apricots in mid-summer, while apples, quinces and pears had to be preserved before the first frosts in November so that they could be served through the winter months. In one particularly hot and fruitful summer, Lady Fettiplace's contemporary Lady Margaret Gardiner sent her apologies for not writing to her friend Sir Ralph Verney herself, since she was 'almost melted with the double heat of the weather and her hotter employment, because the fruit is suddenly ripe & so she is busie preserving'.[32]

The key ingredient in all these fruity preserves was, of course, sugar. Kept under lock and key away from untrustworthy servants and sweet-toothed thieves, sugar was a hugely costly material in medieval and early modern Britain. In the early part of the seventeenth century it was largely imported from the islands off North Africa, such as the Canary Islands and the Madeira Archipelago, where it had been introduced by Spanish and Portuguese explorers. In those days, sugar was not sold loose in bags as happens today, but in cones or loaves (see Plate 5). These were made from cane juice in refineries, the first British examples of which began production in London in the 1540s, appearing across the country by the end of the seventeenth century. Sugar gradually became cheaper over the course of the early modern period as the transatlantic slave trade grew to be a key part of European economies. The first British sugar plantation, which soon relied on the brutal forced labour of people primarily from western Africa, was set up to satisfy England's sweet tooth in the Caribbean colony of Barbados in the 1640s.[33] In England, the appeal of sugary preserves was gustatory and visual, but they were also valued as an important means of preserving expensive fruit that would otherwise go to waste. Cookbook authors Thomas Dawson and Sir Hugh Plat agreed that quince preserved in sugar and water would 'keepe a yeere'.[34] As sugar became more readily available in this period, syrups were used by those lower down the social scale (wealthy farmers, yeomen and citizens of growing towns and cities) as a means of preserving seasonal fruits year-round, though they could not expect to produce the colourful varieties and stupendous quantities that Tudor and Stuart gentry women had access to.[35]

Salt, Pickles, Pies and Cheese

Like sugar, salt was another prized condiment and preservative in the early modern kitchen. As the Elizabethan commentator Henry Buttes wrote in 1599, salt was 'used almost in all meates, to season or preserue'.[36] Since prehistoric times, our ancestors have understood that salt has the power to arrest the decay of foodstuffs, even without knowledge of microbial life. We now know that salt works as a preservative by reducing the amount of water in a food at a cellular level via osmosis, drawing out the moisture which would otherwise fuel bacterial growth. In Tudor England, salt was harvested from mines, like those in Cheshire, as well as being imported from saltpans on the Atlantic and Mediterranean coasts.[37] Today we are perhaps familiar with prosciutto, salami and chorizo, but in the early modern era a much broader range of meat and fish was regularly cured, either with dry salt or with a brine solution of salt and water.

Across Europe, cured fish was a principal part of ordinary people's diets. The vast majority in the sixteenth century relied on preserved salted fish as a means of adhering to the Catholic Church's injunctions to fast on the Fridays, Saturdays and variably Wednesdays of every week, as well as during the long forty-six days of Lent. Over time, this injunction had come to mean abstaining from the flesh of warm-blooded animals and animal products like milk and eggs, so that fish and other cold-blooded animals (including snails, frogs and even turtles) were permitted and consumed. In England, these rules had been relaxed under Henry VIII, and they were further eroded over the course of the Protestant Reformation, when reformers argued that there was no biblical basis for dietary restrictions.[38] However, even after the Protestant Queen Elizabeth succeeded to the throne after her Catholic half-sister Mary in 1558, and in fact right up to the mid-seventeenth century, the English crown continued to enforce fish consumption on Fridays, Saturdays and during Lent as a means of protecting the fishing industry and the navy that then relied on these mariners and their ships. Fasts could in practice be broken by buying a licence. In the parish of St Christopher le Stocks in London, for example, the Lord Mayor paid 6s 8d (roughly the equivalent of £40 today) for a licence to eat meat in the Lent of 1633–4. Laws insisted, however, that even those who could legitimately consume meat still ate fish on fast days.[39] In coastal towns, residents could expect (unsurprisingly) to eat more fresh fish, but in practice these orders meant that many in England continued to eat hard dried fish preserved in salt and packed tightly in barrels to survive its long journey from sea to plate.

The rather nondescript 'saltfish', as it was called, could be salted ling or cod. Meanwhile, preserved herring was the main catch prepared at the centre of England's fishing industry at the port of Great Yarmouth, on Norfolk's east coast. Its flesh transformed from white to red in the process of being salted and smoked, red herring was humorously dubbed the 'king of fishes' by Thomas Nash in his 1599 pamphlet *Lenten Stuff*. It was one of the 'standing provisions' of poorer and richer households alike, and supported a whole industry made up of carpenters, net-weavers, fishermen, salt-house workers and those who washed, packed and transported the final product.[40] Just as they had access to more fresh meat than those lower down the social scale, the elites could also choose to eat expensive fresh sea fish like salmon, trout and mackerel, and they had rights to fish in freshwater ponds and lakes, finding pike, eel, lamprey, tench and more. In the noble Althorp Estate of the Spencer family, in land-locked Northamptonshire, carp, trout and salmon appear with some regularity in the household account books, but dried salted cod (haberdine) was still routinely bought each Friday, Saturday and during Lent to feed a busy working household.[41]

Before consumption, salted fish would have been soaked in water to soften the flesh and release some of the salt. Even so, the monotony of a saltfish diet, especially during Lent, was a long-standing complaint in contemporary diaries and tracts. From the fifteenth century we have preserved the lament of one English schoolboy, jotted down in a private notebook:

> Thou wyll not believe how wery I am off fysshe, and how moch I desir that flesch wer cum in ageyn. for I have ete none other but salt fysh this Lent, and it hathe engendryde so moch flewme within me.[42]

Like Nash's smoked red herring, salted fish and other salted meat could be put through a further drying process to more effectively preserve it. In his tome of agricultural poetry, Thomas Tusser recommended drying out cod: 'Let Haberden lye,/ in pease strawe drye'.[43] Filled with easy-to-remember rhymes, Tusser's *A Hundreth Good Pointes of Husbandrie* was so popular that it was expanded into *Five Hundreth Pointes of Good Husbandrie* in 1573, and remained a key informative text for rural populations throughout the sixteenth century who each day worked to navigate the seasonal availability of crops and livestock. In December, he continued:

> Both saltfish & lingfish (if any ye haue,)
> through shifting & drying, from rotting go saue.
> Least winter with moistnes, do make it relent:
> and put it in hazard, er ere it be spent.[44]

Sausage, which was made of salted minced beef or bacon, could be smoked by hanging it in the chimney. Here 'he may drie', for 'a moneth or two before you take him downe', according to cookbook author Thomas Dawson.[45] As well as its drying effect caused by heat, compounds in smoke are toxic to microbes and delay the oxidisation of fat.

Summer fruits were also preserved through the winter by dehydrating them. If you wanted to keep 'Cherries all the yeare to haue them at Christmasse', one 1608 cookery book maintained, you must first 'be sure that they bee not bruised, and take them and rub them with a linnen clothe'. They would then be transferred to a barrel, laid out with hay in between each row, and left to dry out: 'the warmer they are, the better'.[46] Sliced plums, too, could be dried out to become prune-like in the 'hote sunne', though the cool British weather was not often powerful enough to blast out all the moisture, leading the wealthy to use an oven instead, a piece of equipment beyond the reach of most in this era.[47]

Let's return to meat. Perhaps, like me, you think the crust is the best part of a pie, but throughout the medieval period the thick and rather bland rye flour casing was not eaten, its primary function being to preserve the meaty contents within. Pies could be a substantial size and appear on elite feasting tables, like the ones featured in Pieter Cornelisz van Rijck's kitchen scene of 1610 (Plate 6); those working in towns bought them to take away at cookshops; and they made a hearty dish in taverns and inns alongside a flagon of ale or two. Acting as a convenient means of transporting a meal without the risk of spillage and spoilage, pies were a common sixteenth- and seventeenth-century gift for this reason when friends and family could be reached via horse-drawn journeys over bumpy primitive roads. To make a pie, meat was beaten into a paste in a mortar, seasoned and larded by adding pork fat in order to keep the meat moist. Once the meat had been baked over several hours in the 'coffin', as the pastry casing was known, butter was poured into any remaining pockets of air, and the lid was replaced to act as a seal. First, however, the butter had to be clarified, a process in which the milk solids were removed through boiling (and any scum rising to the top skimmed off), leaving the pure liquid butterfat, the preservative. To make sure that 'your Pies will keepe a great while', Thomas Dawson's cookbook of 1587 directed the reader to parboil the meat first, to fill the pie with clarified butter, and afterwards to use vinegar to further seal the contents and add more flavour during a second spell in the oven:

> First perboile your flesh and presse it, and when it is pressed, season it with pepper & salt whilest it is hot, then larde it, make your Paste of Rye flower: it must be verie thick, or else it will not hold, when it is seasoned and larded lay it in your Pye, then cast on it before you close it a good deale of Cloues and Mace beaten small, and throw vpon that a good deale of Butter, and so close it vp: you must leaue a hole in the top of the lid, & when it hath stand two houres in the Ouen, you must fill it as full of vineger as you can, and then stoppe the hole as close as you can with paste, and then set it into the Ouen again.[48]

Over the course of the sixteenth century, the pie crust was gradually replaced by an earthenware pot, in a process that came to be known as 'potting'. Commonplace in the eighteenth century, potting also involved chopping and cooking meat and sealing it with clarified butter. In pie or pot, meat preserved in this way might generously last a month or two, and emerged from its container as a kind of seasoned pâté.

Pig meat was also cured in salty brines or pickles. A recipe 'to sowce a Pigge', from a 1615 cookery book suggested making 'collars' or rolls of meat, tightly sewing it into a piece of cloth, boiling it and letting it sit in a salt-water solution. This water with white wine, ginger, nutmeg, pepper and bay leaf became a 'sowce' that the meat would supposedly keep well in for 'halfe a yeare'.[49]

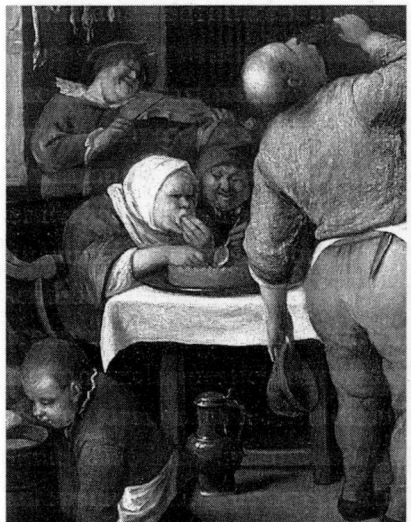

Figure 4. Details from Jan Steen's *The Fat Kitchen* (l) and *The Dancing Couple* (r), showing the filling being eaten from the pie crust with spoons or fingers. Dutch oil on panel, 1650–5 and oil on canvas, 1663 respectively.

Vegetables were commonly pickled. Plat's 1602 recipe 'to preserue Cowcumbers' – what we would today call cucumbers – involved making a pickle of verjuice, water and salt, which was seasoned with green fennel or dill.[50] This recipe might come as a surprise to those aware of the cucumber's dangerous reputation in early modern England. The famous London diarist Samuel Pepys even reported that it was 'eating cowcumbers' that had led to the death of a certain Mr Newburne.[51] This aversion comes from the contemporary understanding of Galenic theory, in which each food and each body was made up of different variations of the four humours: blood, yellow bile, black bile and phlegm, which were hot and moist, hot and dry, cold and dry, and cold and moist respectively. According to humoral theory, cucumbers were too cold and moist to be healthy (which is, incidentally, where we get the phrase 'as cool as a cucumber'), and could lead a person to become melancholic in disposition. This worrying excess could be amended, however, by adding salt to extract some of the moisture.[52] Indeed, given the prevalence of pickled cucumbers, or gherkins, in contemporary recipe books, we can safely assume cucumbers were preserved, eaten and enjoyed. Despite the popular myth, the Tudors – even the wealthy – did consume vegetables, and pickling was a hugely popular method of preserving a whole host of them. As in Plat's recipe, the key preservative ingredient was verjuice, a highly acidic liquid that kills most bacteria, made by pressing unripe grapes or crab apples collected in the autumn. The modern cook might use vinegar in a similar way, and John Evelyn, author of the first English book on salads published in 1699, noted that his wife preferred gherkins made with a wine and alegar (from sour ale or malt) pickle that had been poured boiling hot over the cucumbers once a day for three days. This, he said, 'keepes them long without changing and eate very excellently'.[53] By leaving the cucumbers for a few days before committing them to storage, Evelyn's wife encouraged the cucumbers to start fermenting, a process that did indeed mean – Lady Fettiplace concurred – that they would last longer.[54] The liquid left over from pickled foods, including mushrooms, oysters and anchovies, could then be added to gravies and sauces, or act as a relish.[55]

The final method of food preservation that we need to explore here is cheese-making. While men in rural communities were in charge of ploughing and shearing corn, the dairy was seen as the woman's domain. Work here was among the best-paid that female servants could take on, and making cheese was an essential skill for housewives of all but the highest social rank. As Thomas Tusser said:

But huswives, that learne not to make their owne cheese;
With trusting of others, have this for their fees.
Their milke slapt in corners, their creame all so sost [soured];
Their milk panes so soltie, that their cheese be lost.[56]

Milk made into hard cheese was an important feature in the diets of poorer people, who ate less meat. Paired with a nice chunk of bread, it made an excellent lunch for a working man toiling in the fields, and because it keeps well it was a good product to sell at the local market. Cheese was also an important feature of wealthier tables; softer types of cheese not intended to last for a long time could be seasoned with those elite favourites, rosewater, sugar and ginger.[57] The cheese-making season was best in late spring and summer when milk supplies were at their most plentiful, while butter (which, if heavily salted, could also be sold on in the market) was more appropriately made earlier in the spring; as we've noted, cows would not uniformly produce milk over the autumn and winter. Broadly speaking, women made cheese by combining milk with rennet, which in those days came directly from scraps of a calf's stomach (unlike the rather clinical liquid rennet you can buy today). Left in the warm for at least an hour, the rennet would divide the milk into solid curds and liquid whey. The curds were then broken into small pieces, salted and packed into a cheese mould. Finally, the cheese was wiped in brine – the acidity of which worked to prevent spoilage – dried, and laid up on wooden shelves, being turned every other day or so.[58] Though the whey was a waste product of the cheese-making process, drained from the curds, it was not to be discarded. Fermented until sour and perhaps flavoured with herbs, whey could be turned into a summer drink called Whigge, which was popular among the labouring classes.[59*]

---

* A similar drink called 'blaand' or 'bland' was known to rural workers in the Shetland Islands and in parts of Scandinavia. In 1774 it was described as 'the serum of Buttermilk separated by heat' and 'kept till it is old and sharp'. Whey or buttermilk was mixed with water and left to ferment. George Low, *A Tour Through the Islands of Orkney and Schetland* (Kirkwell, Orkney, 1879), p. 104.

The Pious Kitchen

In these many and creative ways, no one in Tudor or Stuart England would have eaten a diet devoid of preserved foodstuffs. But thriftiness in the kitchen also meant avoiding leaving leftovers on your plate. Like today, wasting food was an issue deeply laden with moral implications. During the sixteenth-century Reformation, when England slowly transitioned from a Catholic country under the authority of the pope to a Protestant state under the crown, a religious mind-set underlined all aspects of daily life. While in the present day food waste is in the mainstream presented in secular terms as an affront to environmental sustainability, writers of the early modern period saw food as a gift from God, so that to waste it was to abuse and reject God's heavenly benevolence. Since the body withered away without it, people believed food physically sustained earthly life by literally replenishing the flesh that was used up day-to-day.[60] And it was food that Christ had chosen to represent his body and blood in the sacrament of the Eucharist in the form of bread and wine. It is perhaps unsurprising, then, that in his *Treatise of Faith* published in 1631, the schoolmaster and Puritan preacher John Ball described the daily rituals of dinner time as a spiritual experience: 'when we sit downe to meate [food], we come to a lively sermon of Gods bountie and love'. The food on the plate, 'is not ours', he argued, 'but the Lords, all the provision are gifts of his mercie in Iesus Christ'.[61] Any food left over, he continued, must not be lost, but we 'reserueth it for good vse'.[62] Another Puritan churchman, Richard Baxter, agreed, writing: 'We must see that nothing of any use be lost through satiety, negligence, or contempt; for the smallest part is of God's gifts and talents, given us, not to cast away, but to use as he would have us.'[63]

This was a lesson repeatedly taught to both children and adults through the parable of the Feeding of the Five Thousand, which is narrated in all four gospels of the New Testament. Many of us are still familiar with the story: a mass of people gathered around Jesus when he went to mourn the loss of John the Baptist, but the group's provisions were meagre, comprising only five loaves and two fish. After giving thanks, Jesus broke the bread and fish into pieces, and by a miracle there was suddenly enough food to satisfy the entire 5,000-strong crowd. Importantly, as recounted in numerous sermons of the early modern period, Jesus then told the disciples to 'gather up the "broken meates"', referring to the leftover food, 'so that nothynge be lost'. The disciples did as ordered, collecting twelve baskets 'ful wyth the broken meates remaining of that which they had eaten'.[64] In his *Of Domesticall Duties* (1622), a manifesto for godly living at home, the clergyman William Gouge used this story to warn servants of their duty to stop food from lying 'in corners moulding, or to be cast up and downe for dogs and cats'. Christ had intended, Gouge argued, for no victual to be lost.[65] Likewise, Robert Cleaver, another Puritan-leaning minister, lectured servants not to waste leftover food, not only because it was 'a hinderance to their masters profit', but because it was 'a great offence to God', taking Jesus' words in the parable as an injunction that should be applied to everyday consumption.[66]

In 1640 the Puritan writer Ezekias Woodward admonished parents of children in the habit of 'spoyling much more than they eat'. With sticky hands and 'mouthes full', children were known (then as now) to leave a fair amount of their food 'spilt on the ground, and some upon the trencher' or plate. Woodward warned that in one family accustomed to such wasteful table manners a 'sicknesse came, and tooke away the parents' as a divine punishment which resulted in the parents' friends having to take on the guardianship of the orphaned children. Woodward counselled that to avoid such retribution parents should teach their children the value of food by recounting histories of famine and starvation to them. Children would surely be shocked into less wasteful eating habits, he reasoned, if they were aware that in the time of King Edward II, in 1315, the kingdom suffered such hunger that people resorted to eating horse flesh… and even their own children. 'Parents or Governours, will take care', Woodward concluded sombrely, 'that there be an humble, thankfull, sober temperate use' of food in their household, and that any 'broken meat be taken up, that the least crum, which can be saved, be not lost; no, not a crum.'[67]

So what was a good Christian meant to do with leftovers if they were fortunate enough to have food to spare? According to the Puritan theologian John Downame in another guide to godly living, anything that remained after a meal should be offered to the poor.[68] For the wealthy, wooden plates or trenchers* could even be painted with intricate biblical scenes and verses that reminded the consumer of this religious imperative once they had had their fill of delicate sweet marmalades and fruit preserves. One trencher (Plate 7) depicts the parable of Dives (meaning 'rich man' in Latin) and Lazarus, the beggar who is shown sitting on the steps to the right of a well-furnished table. According to the gospel of Luke, Lazarus longed to eat the scraps from Dives' lavish meal, but was denied. Dives as a glutton and miser went to hell, and Lazarus ascended to heaven.[69] The story made frequent appearances in the art of the time. While the viewer would delight in the gustatory bounty of the rich household in Pieter Cornelisz van Rijck's kitchen scene, the artist included a warning against wasteful excess by depicting Lazarus on the floor in the far background of the image (see Plate 6).

In rhymes, Thomas Tusser's *Five Hundreth Pointes of Good Husbandrie*, the biggest selling book of poetry in the Elizabethan era, repeated this message. Whereas bad household management would lead to hunger, good management could help the poor: 'Ill husbandrye eatith/ him selfe out a doore/ good husbandrye meatith, him selfe & the poorer.'[70] Other household management

---

* By contrast, the poorest ate off trenchers made of stale bread.

books were more specific. Gervase Markham's *The English Housewife* of 1615, the most popular cookery book of the era, directed the reader that 'the best vse of buttermilke', the liquid left over from churning butter, was for the housewife 'charitably to bestow it on poore Neighbors' who would otherwise lack enough sustenance. Whey, the leftover liquid from the cheesemaking process, should also be given to the poor to drink as they toiled in the fields in the hot summer, or otherwise fed to the pigs.* In making such charitable donations, Markham suggested that the housewife proved herself to be worthy of divine favour: 'no doubt but she shall finde the profit thereof in a diuine place, as well as in her earthly businesse'.[71]

Of course, during the many years of bad harvests when malnourishment was a real possibility for many, the charitable giving of leftovers was even more imperative, and was encouraged with urgency from the pulpit. In that most barren year of 1596, the privy council urged clergymen to instruct their parishioners to skip supper on Wednesday, Friday and fast day evenings, to avoid 'nedeless waste and riotous consumption', and to donate the food that was saved to those in need.[72] Religious leaders encouraged their followers to give away 'at the least so much to the poore, on thy Fasting day: as thou wouldest have spent in thine owne diet, if thou haddest not fasted that day'.[73]

In grand households, the distribution of leftovers to the poor had long been a more organised affair. An ordinance of 1526 for the reform of daily life in King Henry VIII's court required food scraps and dregs of ale to be distributed to those in poverty. This mandate was taken seriously. It was the duty of the officers of the almonry, 'upon paine of imprisonment', to make sure that 'broken meate [leftover food] and drinke be in no wise lost,

---

* In the present, some (artisan) farmers still continue to feed their pigs whey rather than commercialised, commonly soy-based, pig feed. Prosciutto di Parma ham comes from pigs that are often fed the whey created in Parmigiano Reggiano cheese production.

cast away, or eaten with dogges, or lye abroad in the galleryes and courtes'. The almonry was to place two suitable vessels in the scullery – one for solids, and one for liquids – where anyone residing in the court would immediately send their leftovers. Those who disobeyed would be punished on the third offence with the loss of their allowance, lodging and the right to eat and drink at court, a harsh punishment indeed for those who sought the fame and fortune that being a courtier could bring. It was the duty of the under almoner to oversee the daily distribution of the scraps collected in the vessels to the poor who waited eagerly at the gates.[74] From medieval times, the almoner was also tasked with presenting an alms dish to the table and accepting the first slices of meat as a charitable offering.[75] At the gates of the palace, then, two food worlds collided: the court, with its opulent piles of meats, pies, soups, fish, jellies and sweets; and the desperate, hungry poor.

Attitudes towards charitable giving were changing in the Elizabethan period. As poverty increased with the rising population, failed harvests and enclosure of common farming land, the state intervened to enforce parish support for the so-called 'deserving poor', those who were unable to work on account of illness or old age. Relief for the poor was increasingly bureaucratic and systematised by the state, which levied taxes to support people who struggled to afford sustenance, and funded almshouses to house them, rather than the ad hoc and philanthropic system of the past. Unlike in the Catholic medieval period, when wealthy individuals made charitable offerings of food that were, in theory, available to all those in poverty, by the later sixteenth century the unemployed, 'sturdy beggers' and 'idle wanderers' were often seen as 'undeserving' of such handouts. In the household of Henry Stanley, the Earl of Derby, 'vagrant persons or maisterles men' were excluded from hospitality in 1587.[76]

In the early Tudor period, manor houses regularly fed strangers – poor travellers, labourers and others – in their great halls.

This was especially true in the Christmas period, when Twelfth Night feasts like those imagined by Robert May at the beginning of this chapter took place. Yet, by the Jacobean era, the lord and lady of the manor increasingly withdrew to private quarters to eat. Writing in his older years at the time of the Restoration, May – a conservative and a Catholic – looked back wistfully on the 'golden dayes' of his youth, when 'was Hospitality esteemed, Neighbourhood preserved, the Poor cherished, and God honoured'.[77] Nevertheless, the Earl of Derby's household continued in the 1580s to distribute its discarded food and beer as alms to the people in need whom it deemed worthy,[78] and the religious tradition of offering food remains at the gates of grand houses certainly survived into the late sixteenth century. This was still expected of the gentry Willoughby family at Grimsthorpe Castle in Lincolnshire, for example, who gave a piece of saltfish and red herring to the group of poor people waiting at the gates during Lent 1561.[79] When William Harrison chronicled the food and drink of the English in 1577, he noted that when the nobility had 'taken what it pleaseth them' from the table, 'the rest is reserved, and afterwards sent down to their serving men and waiters, who feed thereon in like sort', with anything left 'being bestowed upon the poor which lie ready at their gates in great numbers to receive the same'.[80]

Working for Scraps

While feeding the poor at the gate established a marked and spatial boundary between the rich and the poor, grand households were not entirely closed off from the rest of society; as Harrison's words above suggest, they required a whole host of workers in order to function. In wealthy estates, much of the food was produced on site, with orchards for fruits, gardens for herbs and vegetables, hives that provided honey, bakeries that churned out daily bread and breweries for beer production. As no food was to be wasted, men who worked as officers ('serjeants'), yeomen and clerks in grand households could also expect to receive leftover food parcels as part of the payment for their labour. When King James I came to the throne of England in 1603, he was given an account of how the royal household was managed under his predecessor Elizabeth I, based on 1601, the forty-third year of her reign. The document explains the tasks and remunerations given to each worker. A 'serjeant' was employed to manage the production of bread in the bake-house, for example, and was charged with preventing 'all the lavish expences, all losses, wastes, and filchinges [stealing], within his office'. For this important work he received 11 pounds, 8 shillings and 2½ pence per year, as well as 16 pence a day in board wages. Though it sounds like a pittance in today's money, this made up a significant wage suitable for someone of his managerial importance in contemporary terms. Importantly, in addition to this monetary payment, the serjeant was also to receive 'the brane of all the meale spent in the great bakehouse for his fee'. Whereas most people in Elizabethan and Jacobean England would have eaten cheaper and darker wholegrain bread, the bran – the edible hull, or outer layer, of the wheat grain – was removed to make the kind of fine white breads (called 'manchets') that suited the tastes of the elites. The bran that was salvaged by the serjeant could be used or sold on to others to feed their livestock.

Likewise, the 'serjeant' who managed the royal cellars received the 'lees of wine spent in the yeare', a mixture which could contain some leftover wine along with the murky sediment of the wasted yeast from the fermentation process, perhaps with some grape skins and stems. A number of thrifty uses were developed for this cloudy waste product, including adding it to fresh wine to impart more flavour.[81] Among the many domestic recipes recorded in *The English Housewife* (1615), Gervase Markham suggested boiling 'thicke' wine lees and wheat bran along with sheep's suet with a handful of watercress. Soaked into a linen cloth and tied around a painful leg or foot, this stodgy mixture, he claimed, would help allay swelling.[82] Contemporary writers also recommended reusing wine lees to feed compost and improve soil quality, as is the practice in modern-day industrial wine production, where the yeast in the lees helps break down organic matter.[83] Moreover, the household accounts of Henry Percy, fifth earl of Northumberland, make clear that leftover lees (presumably with some liquid 'lags' or residues) should be further fermented to make vinegar, which as we have seen was employed to make pickles.[84]

In each of the many departments involved in producing food in the royal household the resultant leftovers and by-products were given to the workers. In the 'Boyling House', where meat was cooked, the so-called 'boylers' who worked there were given 'the dripping of the roste' which was collected in a drip tray beneath a spit that turned in front of the open fire. They also received strips from briskets, the sirloin cut of beef and even 'the grease coming of the draweinge of the beefe out of the leade, being in the kittles [kettles] and pannes'. We can imagine these meat juices were used to make sauces and gravies. Meanwhile, in the scalding house, the workers who scalded the carcasses of animals to facilitate the removal of feathers and hair were given the down and feathers of geese, swans and other poultry. Feathers had a number of household uses, including as pillow stuffing and as pen quills. The scalders also took the 'garbidge', a term then referring to animal innards, which as I've suggested could be made into a variety of dishes.[85] Any residues that were left after the workers had taken their fare were likely licked up by a household's pet cat, a common domestic companion then as now, but who also earned their keep by catching any peckish rats and mice in the kitchens and pantries. In van Rijck's painting (Plate 6) the household cat and dog are posed in a standoff over a morsel of fish.

Outside of the household, the very poor – the wives and

children of the men who worked the fields for little money, the elderly without family to support them in their older years, strangers and wanderers looking for food on their travels – bolstered their meagre provisions with the scraps produced in agricultural labour. Gleaning, sweeping the fields for grain that was left over or had been missed in the harvest (see Plate 8), had long been seen as a customary right in England, which ensured that those who could not otherwise afford grain could make their daily bread. One such woman, named only in the historical records as 'the wyfe of John Robyns', a labourer in Hertfordshire, 'did go into the corne feyld of the parish of Hitchin, to gleane graine as is usuall for all the pore to doe', as a way of sustaining herself and her children.[86]

As we shall see later, gleaning – this time from dumpsters rather than from fields – has re-emerged in recent decades as a way of protesting and reclaiming the waste of the modern food industry. Now part of a politico-ecological movement, the practice has a predecessor in the religious worldview that was universal in medieval and early modern England. While those in power tend to condemn gleaning today, contemporary theologians and politicians celebrated this act based on its biblical precedent. In Ruth's story in the Bible, this dutiful daughter-in-law collected the remnants of the harvest from the field of a kind and wealthy man named Boaz. Ruth's actions were allowed under the Old Testament law, which instructed the landowners: 'And when ye reap the harvest of your land, thou shalt not wholly reap the corners of thy field, neither shalt thou gather the gleanings of thy harvest. And thou shalt not glean thy vineyard, neither shalt thou gather every grape of thy vineyard; thou shalt leave them for the poor and stranger.'[87] Ruth gleaned to help feed her mother-in-law, who had recently lost both her sons and so could no longer support herself. In the seventeenth century, this story continued to act as a source of authority and inspiration. Richard Bernard, a preacher at Batcombe in Somerset, published his commentary on the book of Ruth in 1628, praising the protagonist who 'was not ashamed to goe to gleane'.[88] This ancient practice was said to have been granted by the Lord as a way of caring for the poor, and to remind the rich to keep those less fortunate in their thoughts. As late as the mid-nineteenth century, clergymen in Essex, where gleaning was a particularly common occurrence, preached to the farmers that gleaning was 'the sanction of divine command' and that 'preferring the feeding of your cattle to the feeding of your fellow creatures' was a 'wicked practice'.[89] Until the end of the nineteenth century, some rural parishes rang bells morning and evening to define the gleaning day.[90]

However, as Mrs Robyns in Hitchin found out, the right to glean for leftovers was not always accepted. In the court case of 1603–4 which offers us a fleeting glimpse into her life and others of the same social standing, Mrs Robyns was challenged and then assaulted by the son of the farmer who owned the land on which she attempted to glean. The defendant, Edward Hurste, 'did stryke her suddainlye wyth a pitch-forke, being a sufficient forke to work wyth, on the head and on the loynes untill he brake his forke'. Her injuries were so substantial, as several of Mrs Robyns' female neighbours testified, that she could no longer perform any physical labour at all.[91] Unsurprisingly, authorities of all levels recognised the need to enforce rules on gleaning, so that landowners and gleaners did not come into (bloody) conflict, as well as to prevent the fields becoming inundated with the poor at harvest time. Gleaning had become more controversial and the practice increasingly threatened as farm owners enclosed more land.[92] In August 1612, when the major wheat crop ripened, the governors of Lincoln drew up gleaning regulations, arguing that they had been laid down 'to the end that the poor may have and take the gleanings of the fields as fully as had in ancient time been used, and as in charity and by the ancient custom of this Christian kingdom they ought to have'.[93] The local aldermen established rules to crack down on the poor who gleaned the fields before the harvest had finished, maintaining that gleaning could only take place when the corn had been removed by the workers and between the hours of 8 a.m. and 5 p.m.[94] What's more, according to Richard Bernard, the gleaner had to ask the owner and 'acknowledge it a favour' rather than assuming their right, and only gather 'the scattered eares' rather than cutting the edible part from the wheat themselves.[95] As was the case with the poor at the gates of court and wealthy manor houses, the authorities in the sixteenth and seventeenth centuries were keen to distinguish between people worthy and unworthy of this culinary charity. It was only those who were unable to labour that could legitimately glean leftovers from the fields.

Yet many men heaven knows, that have too much,
Whose humours and whose qualities are such
Rather then to helpe poore folke to broken meate,
Theyle cast it under feet for dogs to eate.

Some servants also are so extreame proud,
Though by their wastres daily are allou'd,
Some certaine scraps to give unto the poore,
Theyle rather tak't and cast it out o'th door.

By this we see, Gods word is disobey'd,
Conscience is dead, and charity is decay'd
Rich men that should the poor mans wants releive,
Will sooner from them take then to them give.

L.P., *By the Directions of the Scriptures* [...]
To the Tune of Ayme Not Too High (1650).

Our exploration of food wastage in the early modern period has revealed a society deeply divided, and strictly structured, by economic class. Kings, queens and nobility at the top of the ladder had access to far more food than they could possibly eat, while the poor rioted over shortages and raided fields for the leftovers of the harvest. However, despite changes in the early modern era that saw a decline of traditional acts of charity and an increase in poverty, we have also seen how a religious worldview lay beneath the continued provision of food scraps for the poor, and a universal aversion to wasting food unnecessarily. As we will discover in later chapters, our modern conversations about reducing food waste are most often related to a heightened environmental consciousness. It is worth noting, however, that today the largest chain of food banks in the UK is run by the Trussell Trust, a Christian charity that relies on support from church networks to distribute leftovers and donations to those in food poverty.[96] Providing another example of a faith-driven initiative to redistribute surplus food is the recent partnership between the Muslim community volunteer group Ismaili CIVIC and FareShare, the UK's biggest charity fighting hunger and food waste.[97] As is the case with virtually all religions, which value food as a divinely bestowed gift necessary for bodily life, Islam condemns wasting food or needlessly indulging in excess. 'Eat of its fruit when it yields, and give its due on the day of its harvest, and do not waste. He does not love the wasteful', states the Qu'ran.[98]

Faced with continued bouts of food insecurity and lacking artificial refrigeration to preserve their food, the ordinary people of the early modern period turned to sermons and household guidebooks for inspiration on how to waste less. Of course, as is the case today, not everyone practised what was preached. The clergyman and scientific writer John Beale wrote during the Anglo-Spanish wars of the 1650s that he was 'really much troubled to see our usuall waste', lamenting that there was 'noe use made of the heads, tongue or insides of sheep', 'noe plums ever dryed, nor peares, except in a few houses'.[99] But in general this was certainly not a throwaway society. Early modern cooks across the social spectrum necessarily relied on a number of ingenious methods of preserving bountiful harvests to keep rumbling stomachs quiet throughout the seasons. Elite women worked to make delicate summer-fruit preserves to serve in the bleaker winter months, and even the royal household, where food was most certainly in abundant supply, had developed a system whereby leftovers could be consumed and used by workers and the poor rather than go to waste.

But what to do with the inevitable inedible bits: apple cores, animal carcasses, putrefying meat and rotting vegetables? The following chapter will take us into the homes and businesses of the past to see how waste from food-making was disposed of or recycled into useful products.

2

MUCKHILLS, SKINS
AND ENTRAILS

To save Potted Birds, that begin to be bad.

I have seen potted Birds which have come a great
Way, often smell so bad, that no Body could bear the
Smell for the Rankness of the Butter, and by managing
them in the following Manner, have made them
as good as ever was eat.

Set a large Sauce-pan of clean Water on the Fire;
when it boils, take off the Butter at the Top, then take the
Fowls out one by one, throw them into that Sauce-pan of
Water Half a Minute, whip it out, and dry it in
a clean Cloth inside and out; so do all till they are
quite done. Scald the Pot clean; when the Birds are quite
cold, season them with Mace, Pepper, and Salt
to your Mind, put them down close in the Pot, and
pour clarified Butter over them.

Hannah Glasse, *The Art of Cookery,*
Made Plain and Easy (1747).

Hannah Glasse's recipe to recover preserved bird flesh that
had begun to rot reminds us that the preparation of meat
for consumption was often a gruesome, filthy and even
dangerous business, which was much more visible to ordinary
people than it is today in the modern food industry.

————————

Figure 5. A Worcester pot used for potting meat and fish, a common
preservation method in the eighteenth century, *c.* 1760.

Today, most of us are distant from the visceral acts of butchery:
our meat is presented to us in neat uniform cuts on the supermarket shelves and wrapped clinically in plastic casings. Even at
the butcher's, we are shielded from the bloody, hairy or feathery
carcasses, snapped necks, beaks, bones and gristle, in ways that
may obscure the obvious fact that meat is the flesh of once-living
animals. We have already seen that in the Tudor and Stuart eras,
fewer animal organs and offal were rejected for human consumption than is generally the case in Western cultures today. But have
you ever wondered what happens to the parts of the animal that
humans don't eat? Around 49 per cent of the weight of live cattle
is made of materials that are not consumed by people, 44 per cent
of pigs, 37 per cent of broilers (chickens grown commercially)
and 57 per cent of most fish species.[1] Today, at registered abattoirs
and butchers, the hides of cows and pigs are first stripped from
the carcasses and sold to the leather industry; hooves and horns
can be removed to make glue. Meanwhile, the edible fats around

the heart and kidneys are separated to be rendered into a frying agent (edible tallow or dripping), which is especially popular in northern fish and chip shops. Gelatine manufacturers might also make use of meat scraps, bones and hides, in the production of jellies and sweets. Any remaining fat, bones, eggshells, blood and other animal substances not intended for human consumption go through a rendering process* in which they are crushed, heated to destroy pathogens and separated in a centrifuge into liquid fat (tallow) and protein meal. Even feathers, a small amount of which may be first recovered for use in pillows and duvets, are most often made into a rich protein meal for farm animal feed. The Foodchain and Biomass Renewables Association (FABRA) processes the vast majority of animal by-products (ABP) in the UK – over 2 million tonnes per year – and its members operate at seventeen locations across the country. Animal by-products are divided into three categories based on quality and depending on their grouping they are sold on once rendered to become fertilisers and (if of the highest quality) biofuels, animal feed, pet food, pharmaceuticals, soaps and cosmetics (yes, you might find animal tallow in your lipsticks and eye make-up!). In some cases, animal by-products are simply incinerated, composted (with further treatment), or sent directly to anaerobic digesters where they are converted into biogas and fertiliser as is food from the wider waste stream. It is usually most cost effective, however, to use renderers to make viable products from the parts of animals that have been discarded. As FABRA state, 'renderers are perhaps the original recyclers and our essential role in utilising ABP is often overlooked in discussions on food and food waste'.[2]

On a non-industrial scale, those who reared their own animals in pre-modern times also made use of as much of the carcass as they could beyond the edible meat. In even the humblest of

---

* Rendering facilities also process the carcasses of diseased and dead livestock as well as catering waste.

medieval and early modern households, rendering took place in the form of boiling animal fat (left over perhaps from the cooking pot) over an open fire for several hours to make tallow. When a rush was dipped into the hot liquid it made a simple candle, which – in the absence of electric light, still centuries away – provided the only light in the dark evenings at the farmhouse. To take another example, a dried animal bladder could be used as a container to cook or store food.[3] If a household was lucky enough to have a pig to slaughter in the autumn, its bladder could also be blown up to the delight of the children, who would use it as a football or balloon, as depicted in Flemish and Dutch paintings of the time (see Plate 9). Less frivolously, feathers from chickens, geese and ducks meant a good night's sleep for those who slept on luxurious mattresses. Feathers were also stuffed into pillows, and made into quills for writing. In turn, a quill offered a useful waterproof tube through which to administer home-made medicines. 'If your left nostril is bleeding, use a quill to pour vinegar into your right ear', instructed one recipe book, or if you were suffering from a migraine you might try using a quill to sniff breastmilk up your nose (being careful to note 'if the medicine be for a man, it must be the milke which a Girle suck's on, if for a woman it must be the milke which a boy sucks').[4] Feathers were also adopted for use in the kitchen art of wealthy housewives. Hannah Woolley, in the first female-authored published cookbook in 1661, described using a feather to paint on saffron-water to colour a sugar-paste 'walnut yellow'.[5] Farmers, of course, relied on selling animal products at markets: not just eggs, cheese and meat, but by-products like feathers and wool, the latter making up a central component of the early modern English economy.

Outside of the home, indeed, organised industries developed to make use of animal by-products. In London, a group of professional tallow makers, for example, came together to form the Worshipful Company of Tallow Chandlers around 1300 (receiving full livery status in 1462). As well as selling tallow candles for use within the home, the company supplied the street lights for the City of London, which were compulsory according to city law by the fifteenth century. To access their raw materials, tallow workers set up their premises near slaughterhouses and butchers. In turn, the soap industry relied on tallow. When it was mixed with lye, an alkaline solution made by leaching water through wood ashes or agricultural refuse, tallow was a key component in the type of soap used to wash clothing.

Leather was produced from the hides of animals slaughtered for their meat in a process called tanning. In the countryside tanners were usually large farmers, but the most significant site for the leather industry was London, where tanners organised their supply of animal hides from the city's butchers.[6] Tanners in rural Kent in the sixteenth century might buy their cattle hides from local breeders but also from the London butchers, who had themselves bought their cattle before slaughter from the Kentish Weald.[7] In terms of manufacturing techniques, the leather trade changed little from the fifteenth century until the end of the eighteenth.[8] Hide was first cleaned: soaked in a lime solution, which loosened up the hairs, and the fat and other residues were then scraped off with a knife. Some tanners preferred to use urine, which breaks down into potent ammonia, for the job. Next came the faecal matter: the hide was bated (softened) in an infusion of dog shit or bird droppings (normally from hens or pigeons), which contain enzymes that dissolve the collagen found in skin.[*] Finally, the hide was soaked in a big pit containing an oak bark and water solution for anywhere between six months and two years (with successive Tudor monarchs enforcing minimum processing times to maintain the quality of the product[9]). Tannin, from which this process gets its name, comes from crushed strips of the oak tree's bark, and works to prevent the disintegration of

---

[*] An Act from Elizabeth I's reign enforced the use of 'culver' (pigeon) or hen droppings only. 5 Eliz. 1, cap. 8.

the skin.[10] After it had been tanned, the leather went to a currier, who would cut, dye and finish it for sale onto shoemakers and other leather craftsmen. A lighter 'dressed' leather was made by repeatedly soaking animal skins in oil, beating them and drying them. The finished dressed leather was more suitable for gloves and other leather clothing items.

Since all clothes were made of natural materials, like leather or wool, the creative world of fashion was tied to the grisly business of butchery, and animal by-products also fed our souls in the form of music. 'The delicious tones with which Paganini drew tears from his audience', Richard B. Grantham mused in 1848, 'were produced by the mere friction of the hair and intestines of animals, stretched on instruments constructed with mechanical skill, used with the inspired knowledge of genius.'[11] Indeed, the violin bow of the famous Italian composer Niccolò Paganini would have been created by stretching horsehair across a piece of wood, while the strings were made from animal entrails. String-makers of the past would forge a contract with a nearby butcher to collect fresh intestines straight after the slaughter (normally of a sheep or goat), and once emptied of their manure, clean them laboriously by scraping off any remaining fat and membranes, salting them to preserve them, twisting them into strands of various thicknesses and drying them.[12] Like Grantham, Shakespeare was struck two hundred and fifty years earlier that something so disgusting could create sounds that so delight and move people: 'Is it not strange that sheeps' guts should hale souls out of men's bodies', muses Benedick in *Much Ado About Nothing*.[13]

As well as enchanting our ears with music, the leftovers from butchery were used in visual art to please our eyes. Hog hair collected from tanners, butchers and cookshops was gathered to form coarse paintbrushes. And, later, from the early eighteenth century, iron from cow's blood (and sometimes other parts of the animal, like hooves and horns) was used to form the strikingly blue pigment Prussian blue.

Animals were essential not just to our diets, then, but to a whole host of trades that supplied our ancestors with light, clothing, recreation and art. In medieval London, tanners who received from butchers cattle hides with their horns and feet still attached could sell on the horn to members of the London Horners Company, who made this raw material into various objects like drinking cups, spoons, boxes and combs.[14] Numerous craftspeople transformed animal bones into dice, pins and knife handles, among other things. And, to add one final example, leather cut-offs or hooves were used to make glue. While we may marvel at the ingenious ways in which people of the past made use of the whole carcass of an animal, the meat industry and its offshoots were certainly not waste free. In fact, as I will demonstrate, the unsanitary dumping of discards from the food industries drove the development of public health initiatives as authorities intervened to protect the health and living conditions of their citizens.

Pigs and Pudding-Pits

In 1552 William Shakespeare's father, John, was fined for throwing waste into a 'midden heap' that had accumulated on Henley Street in Stratford-upon-Avon. John Shakespeare was a whittawer (a curer of white leather, which turned white when alum and salt were added) and a glover (a glove maker), professions that along with butchers were notorious for polluting streets with scrapings of fat from animal skins, blood and other leftover animal products.[15] Residents of Stratford were required to send their waste to designated muckhills away from their homes and businesses, but fines like this were not uncommon here and elsewhere across the country. In Cambridge, a pamphlet of 1635 lamented that people were allowing 'the open streets, or under Colledge-walls, Church-walls, or other lanes within the Town' to become strewn with 'their muck, mire, dung, dust, and other filth'. As well as requiring inhabitants to clean around their houses twice a week, the orders set by the town and university stipulated that butchers must not allow any blood to 'runne or come in any streets, lanes, or chanels' on the threat of a 10-shilling fine. 'Paunches, guts, filth, entralls, and bloud' should all be sent to the so-called 'pudding-pits' away from the town at the valley beyond Castle Hill, the orders maintained.[16] While perhaps, as it does for me, the term 'pudding pits' conjures up a fantastical Willy Wonkeresque image – a huge vat of delicious desserts, laden with whipped cream complete with a cherry on top – the word 'pudding' actually comes from the French *boudin,* meaning 'blood sausage', and referred to animal entrails.

As John Shakespeare's case suggests, municipal authorities fought a relentless battle to keep the streets clean of food slops and animal innards, and to banish the putrid smells they emitted to beyond the town walls. Market sellers who left behind decaying leaves, discarded roots, or rotting fruits could also find themselves facing fines in courts for causing a public nuisance with their refuse. In London, waste management systems began in the early thirteenth century, when local orders told inhabitants to keep the immediate area outside of their houses clean rather than throwing their food and bodily wastes out of windows, and the city employed men to sweep up dirt left on streets to designated dumps. By the seventeenth century, refuse was collected by the parish 'scavenger', and from 1671, 'carmen' were employed to sweep the streets and take any household rubbish to the muckhill outside the city, with collections between 5 a.m. and 9 a.m. in the winter and between 4 a.m. and 7 a.m. in the summer.[17] Waste produced in urban environments could then in theory be reused along with manure – the main fertiliser until the early nineteenth century – to help develop the neighbouring farmland.

Outside towns and cities, country dwellers kept their own muckhills. Once it had decomposed given time, heat and moisture, a mixture of excrement and food waste was used as compost on the fields. According to one account of 1688, this 'muck' or 'manure' was made of 'ashes, lime, malt dust, horn shavings, soot', the dung of horses, cows, birds, poultry, 'mans excrements', 'Chaff, Bean Stalks, Leaves of Trees laid to rot, Marl, Chalk' and oyster shells.[18] Writers on husbandry encouraged householders to add brines and beef broth – leftovers from the kitchen that one author maintained were 'commonly throwne away' – to the heap on account of the nourishing salts that they contained. Even soap suds and old rags (made then of course of natural fibres) might also find their way onto the muckhill to break down into useful nutrients.[19] Vegetable waste, malt-dust, entrails and offal, the carcasses of rotten fish, along with blood and dung, were all commonly thrown into 'the dunghill' according to Gervase Markham, writing in the early seventeenth century. A keen husbandman should even seek out the hair scraped from animal hides discarded by the tanners, or the 'waste shavings' of animal horn that horners, tanners and lantern makers found they had no use for, he argued. All these waste materials could be then ploughed, even directly, into the land where they would decay into valuable nutrients over time.[20]

Farmers were yet to understand the role of bacteria in decomposing and fermenting such mixtures, but they envisaged the process of transformation to be much like that which they believed occurred in the stomach, when a variety of food was digested altogether to make a nutrient-rich liquid 'chyle'.[21] In the seventeenth century, scientists hoped to unearth the active ingredient in compost, concluding that it was 'vegetative salts' (nitrates) that were universally found in plants and animals. There was still much debate, however, about what these salts actually were, some conflating them with saltpetre (nitre or potassium nitrate), then an essential ingredient in the creation of gunpowder whose nitrogenic qualities had recently been discovered.[22]

Figure 6. Pigs were kept to consume agricultural and kitchen waste, as in this drawing from Richard Bradley, *The Gentleman and Farmer's Guide for the Increase and Improvement of Cattle*, 1732.

Before they could even reach a muckhill, however, much of a kitchen's leftovers would have been gobbled up by the household pig. In Markham's words, the pig was:

> the Husbandman's best Scavenger, and the Huswives most wholsome sink; for his food and living is by that which will else rot in the yard [...] for from the husbandman he taketh pulse, chaff, barn dust, man's ordure, garbage, and the weeds of his yard: And from the huswife her draff, swillings, Whey, washing of tubs, and such like, with which he will live and keep a good state of body, very sufficiently.

As Markham suggested, pigs snaffled up food scraps and would even consume human waste, in turn producing nitrogen-rich manure which would be used to help grow more crops, and, ultimately, plentiful supplies of delicious meat. This symbiotic relationship between pigs and humans developed at least 9,000 years ago when people realised that rotting leftovers and other household waste – which would attract insects and disease if left to decay – tickled the taste buds of wild pigs. In early modern farms, once they had been fed on leftovers and waste, pigs were let out to graze and root around in fields for more scraps, before being returned to their sties.[23]

Swine were a relatively common sight in towns and cities, too, up until the nineteenth century. According to John Stow's *Survey of London*, first published in 1598, the proctors of St Anthony's Hospital received the pigs that were 'unholsome for man's sustenance' from the markets, put a bell around their necks, and then allowed them to eat from the dunghills until they were fat enough to be eaten by those in the hospital.[24] Outside of this traditional religious exception granted to the Hospitallers of St Anthony, allowing your pig to roam hungrily through the urban streets, or feast from shared muckhills, would likely get you in trouble with the local authorities. In seventeenth-century Cambridge, for example, a swine found feeding in the street would cost you a forfeit of 3 shillings and 4 pence.[25] And in London, too, in 1671 the Common Council, which played a key role in regulating life in the city, felt the need to restate a 1562 statute that people should not allow their pigs, goats or poultry to roam in the 'open streate'.[26] Not only were their bodies and excretions extremely smelly, pigs released from their urban sties might damage people and property as they rooted around in lanes, gardens and houses. In fact, owing to their unsightly and fetid sties, at numerous points from the medieval period onwards, certain urban authorities attempted to remove pigs entirely, allowing them within the town walls only for slaughter. In the 1690s, for example, an Act of Parliament tried to repress pigs being kept at all within the paved part of the city of London within 50 yards of any building, but this did little to stop widespread pig-keeping. John Jolly, who worked in Holles Street (in Marylebone) in the 1760s, is just one of a number of butchers whose names appear in trial records in which they were found guilty for producing 'noisome smells' in residential areas by feeding pigs on the offal and entrails of animals that they had slaughtered.[27] As well as butchers, the cookshops, taverns and bakeries of eighteenth-century London all produced food waste that could be fed to pigs that they housed on their premises. So despite attempts to reduce their presence, town pigs, like their country cousins, acted as lucrative dustbins through which kitchen waste was converted into nourishing meat: cured ham, sausages, fried intestines and lard to cook with.

As larger-scale food industries began to develop from the seventeenth century, pig populations grew on their leftovers. 'Chandlers grains', made up of the 'dregs and offall of rendered Tallow' along with any harder skins or lumps which would not melt, were fed to pigs in 'great Cities or Towns' like London and York, according to Markham.[28] In the 1720s, in his *Tour Thro' the*

Whole Island of Great Britain, Daniel Defoe described how the major bacon-producing areas of Wiltshire and Gloucestershire* had built up around the huge number of dairies which made cheese in these counties and sold it across the country. There the hogs were 'fed with the vast quantity of whey, and skim'd milk, which so many farmers have to spare, and which must, otherwise, be thrown away'.[29] Breweries and distilleries, many of which were established in the areas surrounding London in the eighteenth century, also produced waste products that led to large-scale pig ownership. In 1748 a Swedish visitor to England, Pehr Kalm, recorded how around London the distillers often kept between 200 and 600 pigs, feeding them 'with the lees, and any thing that is over from the distillery', including malt, grain and yeast deposits.[30] Cow keepers, who regularly shared land with breweries and distillers or occupied nearby plots, at least in London, could also buy this waste product to feed their cattle, and in turn they dished out the leftovers from milk production to pigs. Once fattened, the hogs fed on waste were sold alive to carcass butchers for a tidy profit. At Thomas Cooke's impressive distillery at Milbank in London, which housed as many as four thousand pigs living on a diet of waste grain, the animals were actually slaughtered, cut and cured on site in a special processing plant so that Cooke could keep a greater share of the profits himself by selling the finished bacon.[31] Pigs were also tied to the production of starch from grain, which was used in the fashion industry to structure and stiffen fabrics from this period. Swedish traveller Kalm continued: 'In the same way, and with the same object, a great number of pigs are kept at starch factories, which are fed and fattened on the refuse of wheat, when the starch is manufactured.'[32]

---

* The famous Wiltshire Cure originated in the eighteenth century.

Old Father Thames

Have I lived to be carried in a basket, like a barrow of butcher's offal, and to be thrown in the Thames? Well, if I be served such another trick, I'll have my brain ta'en out and buttered, and give them to a dog for a new-year's gift.

Falstaff – William Shakespeare, *The Merry Wives of Windsor* (1602).

In cities and towns, where space was limited, a far greater number of people relied on professional butchers to source, slaughter and cut their meat than in the countryside. The Worshipful Company of Butchers, which was set up to have jurisdiction over the trade of all the butchers who worked within the old city mile, was one of London's oldest companies, with its origins in the tenth century.[33] From its earliest manifestation, this group came under fire for throwing the remaining scraps of meat, blood, hair and the animals' dung into the street, an offence to the city authorities as well as to inhabitants' noses.[34] In the fourteenth century, King Edward III had to get involved, sending the butchers of St Nicholas Shambles to work outside of the city boundaries, after complaints of 'the carrying of entrails and offal, &c., through the streets to the river' in a London area known as Bochersbrigge.* The streets were so dirty that reportedly 'no one, by reason of the corruption of the same and filth can hardly venture to abide in his house'.[35] Their banishment was far from permanent, however. The need for fresh meat brought the butchers back into the heart of the city, and complaints about the noisome disposal of meat appear consistently across the following four centuries. The butchers of the same guild in the area around Newgate Street feature in another case from 1422.

---

* Edward III enforced two such bans on the St Nicholas Shambles butchers, first in 1369 and again in 1371.

After having cleaned the animal's entrails for use in sausages, tripe and fat rendering, they threw their gristly leftovers along with animal dung into the lane by the church of St Nicholas Shambles, drawing complaints from the Friars Minors and parishioners who had to wade through the stinking mess in order to reach the church to pray.[36] One can imagine such filth being unconducive to spiritual enlightenment.

According to John Stow, the septuagenarian who published his *Survey of London* in 1598, Rother Lane in Billingsgate had come to be known as 'Pudding Lane', on account of the 'puddinges' (meaning animal innards) that the butchers of Eastcheape discarded from the scalding houses that they kept there. The lane – which was famously where the Great Fire of London would break out on 2 September 1666 – was full of carts that carried the refuse down to the nearby river Thames, where the butchers kept 'dung boates' full of food waste.[37] Authorities accepted that from here, butchers' waste was to be cast into the Thames when the tide began to ebb, allowing it to be carried out to sea. Out of sight, out of mind. In 1402 the butchers of Eastcheap received a licence to build a bridge over the river Thames from the end of Katherine Lane, on which they could construct buildings and throw 'the intestines of beasts, and the filth and offall of the said beasts' into the water 'to carry by little and little at the reflux of the tide'.[38]

Compounded with illegal dumping, such condoned practices would soon lead to complaints about the state of the Thames' waterways. A lament of 1481 bewailed the overwhelmed city wharves: 'at every Ebbe of the water there remain the Intrails of bestes and other filth of Carion of grete substaunce and quantitee'.[39] It was not just animal waste but other 'dung and filth' that was cast into the river, as described and prohibited by an Act of Parliament in 1535. Since Roman times Londoners had relied on the river's tides to wash away human excrement, with public lavatories set up over each of the Thames' tributaries.[40] The small

Fleet river[*] that emptied into the Thames near Blackfriars had once been passable by boats, but by the seventeenth century it had effectively become a sewer into which the butchers who kept pigs nearby threw offal and innards.[41]

As we've seen, it wasn't just the butchers who were involved in processing the carcasses of animals. The tanners, who turned hide into leather, also frequently found themselves the subject of complaints concerning their dumping of foul waste into waterways, as well as the pungent smells produced in their workshops from decaying meat and the dung used to soften hides. The stench emitted from the huge pits or vats that the hides soaked in could easily waft across the neighbourhood on a windy day. As far back as the reign of Edward I in the fourteenth century, Henry Lacy, the Earl of Lincoln, noted that the tributaries in London under Oldborne Bridge and Fleet Bridge that ran into the Thames had been destroyed 'by filth of the Tanners' along with structural changes that had been made to the channels.[42] As a result, from the medieval period, the tanners were increasingly pushed into the suburbs of the city. As François Jarrige and Thomas Le Roux explain in their history of pollutions, across the world many techniques have been employed to produce leather, but 'one feature common to all was odors and the waste released into nearby waterways'.[43]

The disquiet about the state of the Thames reached an apex in the nineteenth century, as a growing population fed sewage directly into the water alongside mounting industrial waste from slaughterhouses and other animal industries. By 1850, the river Thames was thick and dark, treacle-like with dirt, leading to the infamous 'Great Stink' of 1858, when the hot sticky weather during the summer exacerbated the discomfort of local citizens.

---

[*] The River Fleet was officially converted into a sewer in the nineteenth century and is now visible only as a drainage outlet beneath Blackfriars Bridge.

That year the odour emitted from the water was so bad that the Houses of Parliament had to close down, attempts to oust the smell by soaking the drapes in chlorine a marked failure, while Queen Victoria's pleasure cruise along the river was aborted within minutes.[44] A song sheet from the Victorian era depicts a dishevelled 'Old Father Thames', a beloved character – sometimes even described as a local divinity – who had long been associated with the capital's major waterway. In the image, Old Father Thames rises from the depths of the murky waters, declaring 'Here's a mess I am in', as waste from sewers and factories in the background leads to 'white bait looking black', and to the 'ghosts of departed flounders' – only bones and no flesh – floating to the surface of the river. The popular ballad that accompanied the image sheds more light on the situation. To the tune of a traditional nursery rhyme with which we are still familiar, 'What can the matter be?', it went:

> The fishermen idle, in silence are sighing,
> Their nets are all hanging up rotting and drying,
> Their wives in a stow, 'cause there's no fish for frying,
> And all by these and naughty games.
> Dead is the white bait for city feast founders,
> Formerly eels were caught eight or nine pounders,
> Now I've no eels, – nor no dabs,* – nor no flounders
> So pity pray pity old Thames,

Then the chorus:

> Oh dear! my woes are distressing,
> Never till now I am freely confessing,
> Was I poor old Father Thames such a mess in,
> So pity pray pity old Thames.

---

* A type of flatfish.

Figure 7. A song sheet for 'The Lamentation of Old Father Thames:
written & sung by Mr Hudson & Mr Taylor at public dinners',
drawn by William Heath, *c.* 1850.

It was not just the sewers and butchers to blame. Brewers were
also the subject of city orders from at least the sixteenth century,
which aimed to stop them dumping the 'dregges or drosse' left
over from ale or beer production into London's river system.[45]
'Old Father Thames' continued:

I am used by the brewers of London for brewing,
Their porter, – but seldom for boiling or stewing,
The brewers have aided to bring me to ruin,
And sickened me quite by their games,
For if the beer does not turn out to their wishes,
Through vile common sewers it swashes and swishes,
Comes back again poisons both me and the fishes,
So pity pray pity old Thames,

Oh dear my woes, &c.[46]

From Muckhills to Miasma

The disposal of food waste into waterways, streets, pigsties or pudding pits clearly caused offence to eyes and noses, but perhaps more importantly it was also seen as hazardous to health. With their festering piles of innards, animal dung and rotten vegetables, muckhills had long been assumed to be unhealthy places. In an early medical work of 1542, *A Compendyous Regyment or a Dyetary of Helth*, the traveller and physician Andrew Boorde encouraged his readers to avoid setting up a house near to 'stynkynge and putryfyed' ponds and 'corrupt dunghylles', and instead to seek out a site with fresh clean air. Foul-smelling vapours, he argued, infected the air and penetrated into the body, in turn infecting the blood and creating 'many dyseases & infyrmytyes'.[47] Renewed fears of an outbreak of the plague in 1361, when memories of the pus-filled boils, the vomiting, diarrhoea and piled-up corpses authored by the Black Death were still fresh, led to complaints against the butchers of St Nicholas Shambles on account of the 'abominable stench' of their discarded waste. Left to fester in the streets of London, the butchers' offal was blamed for the sickness of nearby inhabitants.[48] Likewise, when plague hit London during Elizabeth I's reign, the St Nicholas Shambles butchers faced new calls to banish their smelly practice to the countryside.[49] And it was in the plague year of 1625 that the vice chancellor of the University of Cambridge and the town mayor Thomas Purchas agreed to enforce an order ensuring all residents and butchers sent their food waste in 'some vessell' to 'convay to the common muckhill on the bakside of the Towne'.[50]

Figure 8. An engraving *c.* 1656 showing a plague doctor of Marseilles, who wore a mask filled with pleasant-smelling herbs to fend off plague, which was thought to be caused by filthy and rotten smells or 'miasma', like that created by butchers and other food industries.

Underlying this advice were the medical teachings of ancient Greek and Roman writers, especially those of the Greek physician Galen (129–c. 216 CE), which were revised in Europe from the eleventh century based on translations of Arabic medical texts. They then formed the basis of the vernacular health advice manuals that appeared in print from the early sixteenth century. According to Galen's followers, the health of the body was affected by six 'Non-Naturals', referring (somewhat confusingly) not to 'unnatural' things, but to things that were external to the body: air, food and drink, motion (exercise) and rest, sleep and waking, repletion and evacuation and emotions or 'passions', as they were called. It made sense, as the Elizabethan physician Thomas Cogan wrote in his book the *Haven of Health*, that 'aire so corrupted, being drawne into our bodies, must of necessitie corrupt our bodies also'.[51] From the lungs the putrid air travelled to the heart, through the veins and arteries, until it had harmed the whole body. The quality of air was so significant that it could not only afflict the body with disease, but define and alter a person's character. We've seen how eating too many humorally cold and moist cucumbers could induce melancholy in a person. Air that was too cold likewise was thought to breed cold humours in the body, leading a person to take on a melancholic disposition. Those in the northern countries of Europe were therefore 'generally dull, heavy', wrote Robert Burton in his tome dedicated to the topic, *The Anatomy of Melancholy* of 1621. Since they were thought to be melancholic in character, witches were said to thrive in the colder northern air. Bad smells produced from decaying food waste might also induce melancholy, if they did not simply infect the person with plague and illness.[52]

London experienced repeated plagues throughout the seventeenth century, including the most serious Great Plague of London in 1665–6. At the end of the century, the author and vocal vegetarian Thomas Tryon argued that it was because of the poor quality of the air that the 'Cities and great Towns are subject to the Pestilence and other Diseases'. The air was chokingly dense with the breaths of too many bodies, he said, of smoke from chimneys, the 'Liquors Flesh and Fish have been boyled in, mixed with other loathsom and filthy Excrements', and of course the blood of slaughtered animals that 'runs through the Kennels' of the streets, expelling 'deadly Smells and pernicious Vapours' and filling 'the Air with Revengeful Spirits'. Tryon singled out the tallow-chandlers, tanners, leather dressers and soap-boilers as dangerous, 'stinking Trades'.[53] What's more, all those who worked in these 'stinking places', he later argued, became 'dull, foul and gross'. The smells caused lasting changes to the very character or composition of a person: 'if this were not so, a Tallow-Candlers Melting-house, and Butchers Slaughter-houses, and the dark, gross, stinking Smells, and the thick, dreadful fumes thereof, could never be made familiar' to them, he reasoned. Tryon maintained that the air was a kind of 'Spirituous substance' that could penetrate the core of people's bodies.[54] Noses and mouths were the obvious gateways through which noxious vapours and disease could breach the body's defences, causing damage to both matter and mind.

The idea that disease could be caused by festering food waste, and the power of the bad air that it created to infiltrate the human body, survived in varying configurations throughout and beyond the early modern period. As late as the 1850s, noxious air or 'miasma' (from an ancient Greek word meaning 'stain' or 'pollution') as it was commonly known, was blamed for the devastating outbreak of cholera in London, which killed tens of thousands of people. The 'miasma theory' of disease took centre stage in the emergent public health movement of the Victorian era. To those seeking the source of the epidemic, the cramped, waste-strewn and stinking streets that housed the huge numbers of urban poor in industrial cities seemed like the obvious culprit. As the leading sanitary reformer Edwin Chadwick simply put it in 1846: 'all smell is disease'.[55] By this time it was thought that the bodies of those who breathed in a miasma, terrifyingly, might even start to decompose, a condition that was named 'zymotic illness' from the Greek word for fermentation.[56] As more and more Londoners were struck down with diarrhoea and vomiting, their skin shrinking and their eyes sinking as they wasted away from dehydration, the banks of the dirty river Thames were identified as a hot-spot for miasmas. The experiments of the physician John Snow at the Broad Street Pump in Soho in 1854 revealed that it was contaminated water itself, drawn and consumed from the Thames, rather than miasmas in the nearby air, that caused the illness. Yet Snow's theory was not widely accepted at the time, and the germ theory of disease, according to which disease is caused by pathogens rather than bad smells, only started to gain traction in the late nineteenth century. We now know, of course, that the great cholera epidemic was caused by a harmful species of bacterium called *vibrio cholerae*, which had indeed infected London's polluted waterways.

Putrefied Bites

Pigs that foraged for scraps in dirty alleys and feasted on urban muckhills were also identified as a health concern in the early modern period. If they were ingesting infected human and dog excrement on their voracious excursions, pigs could become infested with tapeworm or bladderworm, which littered their flesh with small fluid-filled cysts. Butchers that sold this infected, 'measly' pork were subject to fines, over thirty of which were issued in Manchester between 1648 and 1687, for instance.[57] Without artificial refrigeration, it was also imperative that meat was sold quickly after the animal was slaughtered, so as to avoid selling decaying and dangerous flesh to the customer. In early summer freshly slaughtered beef or mutton would begin to spoil after being transported only 25 miles, and there were others who simply noted that 'putrefaction commences in animal substances almost as soon as life is extinct'.[58] It was for this reason that livestock destined for Smithfield, the largest and oldest meat market in London, was transported through the city on the hoof and slaughtered at butchers' yards around Eastcheap, St Nicholas Shambles, or Whitechapel.* Butchers were given time limits in which to sell their meat once it was slaughtered or risk fines for distributing 'stale Victuall[s]'.[59] Once in the butcher's shop, meat that had been preserved by salting or smoking and was left hanging in an open window was exposed to further threats, especially from flies, which could leave behind wiggly maggots that hatched from the eggs that they laid in the flesh. Those who lived in cities with their cramped streets and the increasingly smoky industries had more to worry about. John Evelyn, a founding Fellow of the Royal Society, argued that meat hung in city chimneys was particularly dangerous because it was exposed to excessive 'corrosive' smoke pollution which would 'so *Mummife*, drye up, wast and burn' the meat, 'that it suddainly crumbles way, consumes and comes to nothing'.[60] Meat hung in the fresh air of the countryside was, unsurprisingly, universally considered to be healthier for you.

Consumers also worried that sellers might attempt to pass off rotting flesh as fresh. Butchers were known to disguise old birds by, for example, greasing the skin and plastering on a powder that made the bird a more pleasing colour. Eliza Haywood, who wrote *A Present for a Servant Maid* in 1743, accused the London butchers of 'blowing', artificially inflating the meat carcass with their breath, or even stuffing it with rags to make it appear fuller. She also warned her readers that some fishmongers coated gills with fresh blood to imitate newly caught fish.[61] Butchers might

---

* Smithfield was the site of the city's livestock market as early as the tenth century. By the mid-eighteenth century it was no longer a suburban area as the city expanded. In 1852 the Smithfield Market Removal Act relocated the live cattle part of the market to a new site at Copenhagen Fields in Islington.

Figure 9. Consumers had to know how to spot rotten, diseased or
adulterated meat when they went food shopping. James Pollard,
The Meat Market, 1822.

also purposely leave carcasses filled with blood, which would
make joints appear fresher and plumper, and if sold by weight
would fetch the seller a heftier price. When the inexperienced
housewife returned home to preserve the meat in a brine solution,
however, she would, on account of the excess blood, find it to
'not take salt or long continue sweet', that is, it would quickly
perish.[62] No wonder then, that food writers of the seventeenth
and eighteenth centuries instructed their readers on how to spot
rotten or infected foods, especially meats, at the market. Watch
out for dull, sunken fish eyes, greening guts and the bent or stiff
tails of shrimps, one pamphlet warned. To check the freshness of
dried ham or bacon the savvy shopper should stick a knife under
the bone, being the part inclined to decay the fastest, and smell it.
Failing that, they should look at the fat on the edges of the joint

and if it is white then it is still good, but, if it is turning yellow 'and of a greasy softness', then 'it begins to be rusty there' and is best avoided.[63] Another author advised those intending to buy a pheasant to try looking into the bird's bottom, or 'vent' (cloaca). If 'it be green in the vent [...] then it is stale kill'd', while if 'white in the vent, then she is new kill'd'. The purchaser should also be sure to check for any flour or white powders used to disguise its age, the advice-book continued.[64]

To 'prevent Sickness and Disease', butchers who resorted to such underhand methods could be reported, fined and their meat compounded. In the York Shambles, the traditional location of the butcher's trade,* rotten meat was burned publicly at each Thursday market. In some cases, butchers found guilty of selling poor-quality meat, or meat from an animal that had died of old age or disease rather than being slaughtered specifically for consumption, could be placed in the stocks or pillory as a way of discouraging others from committing such offences. Elsewhere confiscated meat could be fed to dogs or pigs, or, in the case of Richard Whitehead, seller at Newgate Market in December 1763, thrown into the River Thames (ironically further exacerbating the problem of river pollution which was already blamed on butchers).[65] In London, the butchers and fish guilds had long set regulations, enforced inspections and imposed punishments to prevent such abuses.

Once it had been brought into the home, even good-quality meat still risked succumbing to putrefaction or contamination. Through experience, home cooks knew the importance of protecting food from exposure to the air to prevent its decay, but they were not yet familiar with the role of bacteria in this process. Housewives and cooks would often wait for cooked meat to cool down uncovered before potting it, for example,

---

* 'Shambles' likely comes from the Old English word 'shamel', meaning the stalls on which butchers displayed their meat.

allowing for the build-up of harmful bacteria.[66] And even while warning their readers of deceitful butchers who made foods look fresher than they were, cookery and household advice books included recipes for reviving tainted meat at home that would certainly not pass our health and safety tests today. In the mid-eighteenth century, Hannah Glasse's recommendation that started this chapter, to repot birds that had gone off and were covered in rank butter, is one stomach-churning example. In another, over a century earlier, Sir Hugh Plat noted that he'd been 'enformed by a Gentlewoman of good credit' that if venison had started to go bad one must first cut away any green flesh, remove the bones, and 'bury it in a thin olde course cloth a yard deepe in the ground for 12 or 20 houres space'. After this period, he claimed the venison would be 'sweet enough to be eaten' once again.[67] Writing in the Jacobean era, the expert on husbandry Gervase Markham recommended that spoiled venison could be soaked in 'a strong brine' made of ale, wine vinegar and salt, before being pressed, parboiled and roasted.[68]

While the wonders of microbial life were first witnessed under the microscope in the late seventeenth century by the Dutchman Antonie van Leeuwenhoek, most scientists continued to explain fermentation and decay as a spontaneous event that occurred when food encountered air. A century later, the Italian priest Lazzaro Spallanzani set out to prove that it was in fact microorganisms in the air that caused putrefaction. Having boiled a broth, he placed it in a sealed container from which he had removed all the air to create a vacuum. When Spallanzani reopened it after storage he found that the broth had not spoiled. His opponents, however, simply argued he had proved that spontaneous generation required air to work, not that microorganisms had done the spoiling. Only in the 1860s would the French chemist Louis Pasteur finally put the debate to bed with an experiment in which he boiled a meat broth in a 'swan-neck' flask, with a narrow s-shaped tube as its opening. Although air was able to reach the food, no decay occurred, as the microbial particles became trapped in dust and moisture on the inside surfaces of the tube. This breakthrough was important to our understanding of how human disease occurs, sounding the death knell for the miasma theory, as well as of how food putrefies. It was not the air itself that caused spoilage, but the tiny organisms living within it. Even without comprehending this, early modern shoppers and cooks knew to look out for decaying or infected meat that risked making dinner guests unwell, and many would have recognised that despite the tricks suggested by contemporary cookbook authors, little could be done to rescue meat that had truly begun to turn.

What to do with rotting leftovers was a question that concerned local authorities as well as residents. As we have seen, butchers and the many industries that developed to exploit animal carcasses into useful products, like the leather tanners and tallow chandlers, were most notorious for dumping smelly detritus in the streets and rivers. As towns and cities grew, these industries were increasingly forced into suburban areas where they caused less offence. In the City of London, for instance, the sixteenth-century rise in population meant that by 1619 there were just forty men employed in the leather industry but as many as 3,000 working in the suburbs.[69] Since miasmas were then thought to be the cause of disease, orders to banish slaughterhouses from city centres to the outskirts were an instinctive reaction to plague outbreaks from the medieval period until the end of the eighteenth century. Yet they were unsuccessful in keeping butchers' waste away from the inhabitants for substantial stretches of time. Londoners in the 1840s were still complaining about the overpowering stench caused by the streams of filth disseminated by city slaughterhouses, much as they had two hundred years previously. In Victorian Bristol, too, slaughterhouses continued to be 'scattered over the town without any special regulations'. A shopkeeper whose lodgings overlooked one slaughterhouse reported that dead pigs were left so long before they were cut up that he'd seen 'maggots fairly dropping out of them', while their entrails were thrown onto the dunghill. The smell was so bad that he shut all his doors and windows, shoving bits of cloth into the key-holes to try to prevent the noxious air from entering his living quarters.[70]

It was in the nineteenth century that public health initiatives, in which the government took responsibility for the health of the population, began to be introduced on a substantive scale. As germ theory replaced the miasma theory in our understanding of how food decays and disease spreads, sewage systems, the regulation of industrial waste and food waste disposal all came under increasing governmental regulation. The 1848 Public Health Act was the first distinct law on public health in England and Wales. It established a Central Board and Local Boards of Health with authority over the establishment of slaughterhouses, and 'offensive trades' like blood, bone, tripe and soap 'boilers' and 'tallow-melters'. An 'Inspector of Nuisances' was empowered to inspect shops or slaughterhouses selling meat and fish, and to fine them if food was found to be unfit for consumption. Voluntarily, the Local Board of Health could provide vessels for householders to collect their waste so that it could be removed, echoing orders made on a local scale in seventeenth-century London. The Public Health Act of 1875 consolidated this and other laws concerning sanitation, and was more universally enforced.

Though pigs had long acted as energetic food waste recycling machines, the sanitary reformers of the start of the nineteenth century saw them as dirty, legislating fines for keeping pigs in the crowded houses of the urban working classes where humans and pigs often lived cheek by jowl. The reform efforts of one impassioned Medical Officer of Health, for instance, cleaned out all the pigs from Kensington in London by 1878, no small feat given the 'shanty town' there had housed 3,000 animals in only 9 acres of space.[71]

The Town Police Clauses Act of 1847 had legislated against throwing food waste from households and food manufacturers into the streets.[72] At the same time, motivated by a desire to clear the congested city streets of livestock and slaughterhouse refuse, the slaughterhouse reform movement began to campaign for the private slaughterhouses, which were owned by butchers themselves and often situated beneath their shops, to be replaced with public abattoirs. As well as condemning the poor conditions in which animals were routinely slaughtered in private establishments, Victorian campaigners highlighted the inadequacy of state regulation over butchery, which resulted in blood and other by-products flowing aimlessly into sewers or being swept onto muckhills rather than being reused in some profitable way. Public abattoirs owned by the local government appeared in Norwich, Liverpool, Manchester and Leeds in the late 1800s, though London and Bristol were slower to adapt.* London butchers complained that moving slaughter outside the city would cause a meat shortage, because it would rot before it had reached the customer.[73] Sanitary reformers in turn complained that London's streets were 'flowing in streams of blood and filth in all directions', with hordes of pigs 'wandering and feeding upon all the offal and blood that they meet with'.[74] Only in the early twentieth century had the accumulative work of various sanitary reform movements successfully banished most of the food and animal by-product industries from the urban environment.

In the twenty-first century, the filthy smells and pollution caused by slaughtering animals and the rendering of their bodily by-products are kept hidden from public view behind the closed doors of large factories and plants. We no longer need to consult

---

* Manchester opened its public abattoir in 1872. Birmingham's dated from 1895 while Leeds had a public abattoir by 1898.

a manual to learn how to recognise 'blown' or 'measly' meat at the butcher's or supermarket. Instead, the government's Food Standards Agency imposes strict rules on all businesses that handle meat at any stage in its production, in order to protect consumers from diseased or decayed products. Top-down public health initiatives also control the reuse of leftovers from an animal's carcass. Following an outbreak of foot-and-mouth disease on a UK farm in 2001, caused by the illegal feeding of untreated food waste to pigs, the practice of feeding animal remains in the form of processed animal protein (PAP) from renderers to farm animals was banned across the EU. This extended earlier bans on PAP being fed to ruminants (cattle, goats, sheep etc.), following the BSE, or mad cow disease, outbreak of the 1980s and 90s. Only in August 2021 did the EU in part lift this legislation, allowing pigs to be fed the leftovers from poultry carcasses, and increasing the possibility of rearing pigs on food waste once again, a move broadly celebrated by food waste activists, but not yet implemented by the UK government (as of 2023).[75] Both the strict controls that govern the functioning of the modern food industry and public concern about its impact on the health of the population have their roots in public health initiatives that began to evolve in the medieval period.

As poor Old Father Thames lamented, our rivers and seas have long been exploited as the final resting place for our grisly leftovers. It was these waterways, however, that offered our ancestors access to the wider world, with all the culinary wonders it had to offer. In the next chapter, we'll follow the routes of some pioneering voyages on the high seas, considering how matters of food waste and preservation fuelled overseas exploration and an increasingly globalised food system.

3

THE WORLD IN A TIN CAN

Biscuit, Sea

A large lump of dough, consisting merely of flower and water, is mixed up together, and placed exactly in the centre of a raised platform, where a man sits upon a machine, called a horse, and literally rides up and down throughout its whole circular direction, till the dough is sufficiently kneaded [...

The first man on the farthest side of a large table moulds the dough, till it has the appearance of muffins, and then delivers them over to the man on the other side of the table, who stamps them on both sides with a mark [...]

The fifth arranges them in the oven [...]

The biscuits thus baked are kept in repositories, which receive warmth from being placed in drying lofts over the ovens, till they are sufficiently dry to be packed into bags [...]

At Deptford the bake-house belonging to the victualling-office has twelve ovens; each of which bakes twenty shoots daily: the quantity of flour used for each shoot is two bushels, or 112 pounds; which baked, produce 102 pounds of biscuit.

William Falconer, *A New Universal Dictionary of the Marine* (1815).

Ship's biscuit, or sea biscuit, was the most common feature of seafarers' diets, which was made on industrial scale for the navy. The hard, dry biscuit was one of the few foods that could last the months at sea. You can make it yourself by simply mixing 450g of flour with enough water to make a stiff dough (*c.* 200ml), divide it into four balls about 1cm thick, prick them with a fork (the little holes help release more moisture), and bake them for two hours on a low oven. Then leave them to dry out further.

———•———

The American Revolution began with food waste. On 16 December 1773, 342 chests of tea belonging to the British East India Company were thrown into Boston Harbour by American colonists in protest at the Company's enforced monopoly over American tea imports and the British government's imposition of taxes on the colony. Disguised in Native American headdresses, around sixty men snuck aboard three British ships left in the harbour to destroy the expensive import, which was supposedly worth over $18,000, or $1–2 million in today's money. The Boston Tea Party left the water murky with tea leaves. British retaliation, harsh economic punishment dished out in the form of the so-called 'Intolerable Acts', had the unforeseen effect of uniting the colonies against British rule. War soon followed, and the Thirteen Colonies were finally declared independent of Britain in 1783. Despite this loss, Britain's overseas empire still grew in the eighteenth century: the British East India Company gained a significant victory in Bengal, India, in 1757, and by the end of the century Britain had founded a colony in Australia, later laying claim to New Zealand. Other European powers likewise sought out wealth, power and food through global empire-building in the eighteenth century. Often described as the 'real first world war', the Seven Years' War had broken out in 1756 as Britain and France sparred over land in North America. In Europe, Spain, Prussia, Saxony and Austria were soon dragged into the war as alliances and territories were violently contested. Following the

French Revolution, the military commander Napoleon Bonaparte seized power in 1799, and quickly initiated further global conflict as he sought to dominate Europe, vying with rival countries for control of seas and land within and outside of the continent. When he was finally defeated in 1815 at the Battle of Waterloo, Britain's navy emerged as globally dominant, setting a clear path towards even more colonial acquisitions over the course of the nineteenth century. Food waste certainly didn't gel with imperial ambitions. How best to effectively feed growing armies, navies and populations of settlers as they travelled far from home was a pressing question for Europeans in this period of war and empire-building.

As the British expanded outwards, politicians and philosophers alike sought out new and efficient ways to feed the expanding population living on their own soil. The Board of Agriculture was set up in 1793 to research into and embolden the development of so-called 'enlightened' British husbandry. Prizes were offered by both governments and organisations for new agricultural inventions and improvements in the size and nutritional properties of crops that reduced waste and increased outputs. Earlier in the century, the introduction of the Norfolk four-course rotation system, first invented by farmers in Flanders, had brought about enormous improvement in agricultural efficiency in Britain. Crop rotation, when a sequence of different types of crops are grown in the same field, improves the soil quality and prevents the appearance of pests and weeds. The four-course system negated the need for a traditional 'fallow year', a break in cultivation when an area of land was left empty to allow it to recover its nutrients after successive harvests. In the new system, which came to be associated with the statesman Charles Townshend, who promoted it from his Norfolk estate in the first decades of the eighteenth century, a grazing crop (clover) and a fodder crop (turnips) for ruminants were added to the rotation in each field. Clover provides a vital plant nutrient by fixing atmospheric nitrogen in the soil, while turnips smother weeds. 'Turnip' Townshend, as he was nicknamed, was among those who recognised that since more animals could be fed on the land using this system, it would also provide more milk, cheese and meat to feed people, as well as manure to further nourish the soil. New technologies were concurrently making agricultural work more efficient and less labour-intensive. In the 1730s, Joseph Foljambe's patented Rotherham Swing plough removed the need for an ox driver so that ploughing could be carried out by one man alone for the first time. That decade, Jethro Tull also published an account of his waste-saving invention, a mechanised seed drill that required 30 per cent fewer seeds to be planted than when they were sown by hand. By the end of the century, the threshing machine was revolutionising farming, taking over the laborious manual job of separating grains from the stalks and husks, and saving around 70 per cent of labour.[1] As small farms grew into larger, higher-yielding enterprises, farming slowly transitioned from a subsistence activity that supported families with food and resources to a capitalist one in which excess product was sold on for a monetary profit.

Even animals were reimagined in this new era of agricultural efficiency, with the development of selective breeding, a way of creating livestock whose bodies produced more – and wasted less – food. From his ancestral land, Dishley Farm near Loughborough, the agriculturalist Robert Bakewell bred a new type of sheep called the (Dishley) Leicester. Raised to the status of a 'celebrity' in its own time, this novel breed fattened faster and had a lower ratio of bones to meat. As one contemporary commentator remarked, 'the improved breed has a decided preference: for surely while mankind continue to eat flesh and throw away bone, the former must be, to the consumer at least, the more valuable'.[2] It was also in the eighteenth century that English pig farmers began importing Chinese swine and interbreeding them with the spindlier European examples (see Figure

6) to create the fatter and faster-growing breeds that are more familiar to us today.

Fed by the wealth and resources pilfered from Britain's growing empire, the Industrial Revolution was forged alongside the Agricultural Revolution. As agricultural production became more efficient, there was less work for agricultural workers, and more people flooded into urban environments. London grew from around 300,000 inhabitants in the sixteenth century to 800,000 people by 1800.[3] This was an age of technological innovation and invention: gas lighting, powered by coal, was first developed in 1792 by William Murdoch, and would soon replace animal tallow candles and oil lamps to light the streets. Food production was increasingly mechanised as the Industrial Revolution advanced. From the 1780s, Matthew Boulton's and James Watt's steam engine, an improvement on the model introduced earlier in the century by Thomas Newcomen, was slowly beginning to be used in the manufacture of beer in place of the horses that had supplied the power to grind malt.

Agricultural and industrial change would soon dramatically alter both local and global food supply chains. For most of the eighteenth century, food was transported in wagons and carts using the power of animals, just as it always had been. In the early years of the century, Daniel Defoe, author of *Robinson Crusoe*, described how roads that connected the trading centres of the country – muddy, sometimes flooded, worn with the hooves of droves of cattle, sheep and hogs – were being improved, with huge benefits to the food industry. According to Defoe, herrings being transported from coastal towns such as Yarmouth and Ipswich 'frequently stink' before their arrival at towns in the Midlands. Yet, if roads were improved, Defoe argued, journey times would be shortened so that fish arrived at their destination in a fresher state and more of the product could be consumed by the inhabitants.[4] As road engineers got to work, over the course of just fifty years between 1780 and 1830 the stagecoach journey between London and Edinburgh had shortened from two weeks to two days.[5] Britain's waterways also received an upgrade in the Georgian era. The Sankey Brook in north-west England was made navigable in 1757, making it the first modern canal, with support from nearby industries, including the salt manufacturers on the River Weaver. Salt was a valuable commodity, used of course in food preservation and cheese-making. Among the principal uses of canals put forward by the pioneering Scottish engineer Thomas Telford was the movement of 'Groceries and Merchant-goods for the Consumption of the District through which the Canal passes', as well as transporting manure for the improvement of local fields.[6]

Most significantly in terms of food supply, by the end of the Georgian period, the railway era was already beckoning. Scottish inventor William Murdoch created the first prototype for a steam locomotive in Birmingham in 1784, Richard Trevithick built the first working steam train to run on smooth rails in 1802, and in 1825 the Stockton & Darlington line opened – the world's first public steam railway. Following the opening of the Liverpool and Manchester Railway in 1830 – the first inter-city service – the railway network grew rapidly, carrying freight as well as passengers. Local products could now be joined on the plate by foods that had travelled across the country before being consumed. The Sandbach and Wheelock branch of the North Staffordshire Railway came to be known as the 'salt line' soon after it opened in 1852, as it carried salt from Cheshire producers to industrial ports like Liverpool, in turn carrying back the coal and equipment needed at the saltworks.[7]

Food was being transported with speed across previously unimaginable distances, not just nationally but globally, in line with the expansion of overseas exploration and empire. To take one example, Defoe wrote of the people of Poole in Dorset who were famed for pickling oysters which were put into barrels and 'sent not only to London, but to the West Indies', parts of which had been colonised by the English since the seventeenth century.[8] The tea destroyed in Boston's harbour at the start of America's revolution in fact originated in China, and would have been sweetened in Britain by consumers rich and poor with sugar grown on the slave plantations in the West Indies. In the eighteenth century India ate up 10 per cent of Britain's exports including preserved salmon and other fish, hams, dried fruits and liqueur, all comestibles that could last the long sea voyage without spoiling.[9] As will become clearer over the course of this chapter, the quest to preserve food and prevent it from going to waste went hand-in-hand with the formation of the modern globally interconnected food system.

The Invention of the Tin Can

A letter from Kensington Palace dated 30 June 1813 reports that Queen Charlotte and her son the Prince Regent (later George IV) had tasted – and enjoyed – canned beef for the first time at a banquet hosted by Prince Frederick, Duke of York. King George III's royal household were so impressed, wrote the secretary to Prince Edward, Duke of Kent, that they wished to help in 'proving the merits of the things for general adoption'.[10] Earlier that year, the recipients of the letter, Bryan Donkin, John Gamble and John Hall, had established the first food-canning factory in the world on Blue Anchor Road, Bermondsey, South London. We can begin the story of the tin can, however, a few years previously in Napoleon's France. As supplies of sugar, brought to Europe from the New World, were cut off during the Napoleonic Wars, the French government sought out new sugar-free ways of preserving food stocks.[11] They were eager to find a way of feeding the vast numbers of men serving in the *grande armée* and the French navy, who served far from home and far from fresh food, offering up a prize of 12,000 francs to anyone who could come up with a new effective preservation method.

The man who would rise to the challenge was Nicholas Appert, an accomplished chef who, having survived the revolutionary Terror of 1793–4, devoted himself to experiments in food preservation from his confectioner's shop in Paris. As Appert put it, traditional preservation techniques relied either on 'dessication', meaning drying out the foodstuff, or 'mingling' it with a foreign substance like sugar, syrups, pickles, or salt. These methods were problematic, he argued, given that they dampened odours, changed tastes and hardened textures. Instead, the technique he developed – heating bottles of food in a basin of water – would preserve the look, taste and smell of the food, as if, almost magically, no time had passed since its creation.[12] The son of an innkeeper and brewer, Appert had grown up 'in the pantries, the breweries, store-houses, and cellars' of his home town of Châlons-sur-Marne in the Champagne region of France, where no doubt he witnessed the workers' relentless struggle to arrest the decay of their food and drink stores. After moving to Ivry-sur-Seine on the outskirts of Paris in 1795, Appert had refined his method enough to send some of the bottles to be tested by the French navy at Brest. Eight years later, a letter sent from the Council of Health to Caffarelli, the Maritime Prefect at Brest, confirmed that even after spending three months in storage before they were used, 'the broth or soup (bouillon) in bottles was good'. So too was the boiled beef in broth (if a little weak), the beef was 'very eatable', while 'the beans and green peas, prepared both with meat and vegetable soup, had all the freshness and flavour of recently gathered vegetables'.[13]

At this point, Louis Pasteur's revolutionary work explaining how heat destroys bacteria in liquids was still around half a century in the future. Yet Appert's method revolved around 'the simple principle of applying heat in a due degree to the several substances, after having deprived them as much as possible of all contact with the external air'. This same idea – that the decay of food was delayed by creating an airtight container – had in fact been routinely adopted by housewives throughout the early modern period. Hannah Woolley's *Queen-Like Closet* of 1670 included a recipe 'to preserve barberries without fire'. After sugar had been layered in between the fruit, the pot was sealed with a wet animal bladder stretched over the top, so 'that no Air get in, and they will keep and be good'. Woolley also described how grapes could be kept fresh by placing them between layers of oats in a box which was closed quickly so that 'no Air get in'.[14] Others made preserves by bottling fruits and immersing them in hot water, and the seventeenth-century scientist Robert Boyle had experimented with heat and steam as ways of manufacturing air-free containers.[15] Appert, however, was the first to establish the bottling (or canning) method on an industrial scale. In 1804, he opened the first food bottling factory in the world from his house in the small town of Massy, south of Paris. Here Appert preserved a huge variety of foods and drinks, supervising the collection of vegetables from the garden of his property, and ordering in meat and milk from the surrounding countryside. A whole range of prepared dishes were also preserved, including a 'fricasse of fowls', and 'a matelot of eels, carp, and pike, with an addition of sweet-bread, mushrooms, onions, butter and anchovies, all dressed in white wine'. In the case of beef, Appert would, after having cooked it until it was three-quarters done, pour the meat and the broth in which it was boiled into a glass jar, which he quickly sealed with a cork. Appert then wrapped the bottles in linen cloth and placed them in a large cauldron filled with water, which he heated for one hour, letting them sit as the water cooled down for a further half an hour, before sealing the cork securely with rosin. 'At the end of a year, and a year and half, the broth and boiled meat were found as good as if made the day they were eaten', he concluded.[16] For any foodstuff his process essentially involved four steps, which he described as follows:

1st.   In inclosing in bottles the substances to be preserved.

2d.   In corking the bottles with the utmost care; for it is chiefly on the corking that the success of the process depends.

3d.   In submitting these inclosed substances to the action of boiling water in a water-bath (Balneum Mariae), for a greater or less length of time, according to their nature, and in the manner pointed out with respect to each several kind of substance.

4th.   In withdrawing the bottles from the water-bath at the period prescribed.[17]

This method won Appert the French Ministry of the Interior's coveted prize on the condition that he made it public. He went on to do just that, publishing a treatise entitled *The Art of Preserving Animal and Vegetable Substances* (*L'Art de conserver, pendant plusieurs années, toutes les substances animales et végétales*) in 1810. On 30 August, just a few months after Appert's method had been published, the London merchant Peter Durand acquired a patent for the same technology, with some additions. Durand laid claim to a wider variety of canisters including those made from pottery and metal as well as Appert's glass, and his patent covered containers fitted with corks, as Appert suggested, but also sealants made from bladders, cloth, leather and other materials. The speed with which Durand acquired the rights suggests that an intermediary had introduced him to Appert's work. Indeed, the academic Norman Cowell's detective-like research has uncovered that another Frenchman – the inventor Phillippe de Girard – had done so around a year previously.[18] Just two years later, Durand would sell his patent to the manufacturers Bryan Donkin, John Hall and John Gamble for the sum of £1,000. In 1813, at their purpose-built factory on Blue Anchor Road in South London, these three men produced the world's first tin cans. The robustness of tin was an obvious advantage over the use of glass, though the new canisters were hefty, weighing anything between 4 and 20 lbs – that's up to twenty times heavier than the tin of baked beans in your kitchen cupboard!

Figure 10. An early example of a tin can used for the preservation of food, made by Bryan Donkin & Co., London, England, 1812.

Donkin, Hall and Gamble's use of tin to preserve food was not entirely novel. In the 1770s, the British-Dutch mercenary Captain John Stedman was employed to quell a revolt of the Africans enslaved on plantations producing sugar, coffee, cocoa and cotton in the Dutch colony of Surinam. In his memoirs he recorded a 'delicate present' of roast beef that had been sent from Europe to the Dutch forces. As he described, the meat was put in a tin box, entirely covered in gravy or dripping, and the tinplate container soldered shut. 'By this means I was told it may be with safety carried round the globe,' Stedman wrote.[19] Tinplate was available to Donkin,

Gamble and Hall in a way that it wasn't for Appert; during the Napoleonic Wars the British had access to a far greater and superior supply of tinplate than the French, Appert later explained.[20] In early examples, the tinplate was cut by hand, bent in the shape of a cylinder, and the edges were soldered. Next a tinplate circle was attached to one end, the tin filled with its cooked meat or vegetable mixture, and another circle was soldered to the other end. The tin was then heated to a high temperature for several hours before the hole that acted as a vent for steam to escape was sealed to prevent any air getting in. Finally, the tins were covered in lead paint to protect them from corrosion, especially from salty sea water, and a ring attached so that they could be carried or hung up on a peg.

From the outset, the tin can was of great interest to the British navy. In November 1812, early samples made by Donkin, Hall and Gamble were sent to Captain George King, who took them on his round trip to Jamaica in the ship the *Mary & Susannah*. The preserved meats and soups survived the hot weather and the long voyage, Captain King and four spare cans returning in good condition in July 1813. The navy soon made orders for tins to supply ships across the British West Indies. Early tins also travelled in British ships sailing to India, large parts of which were under control of the British East India Company; to Saint Helena, an island in the South Atlantic that was also under British East India Company rule; to China with merchants trading in the Far East; and to the Cape of Good Hope at the southernmost tip of Africa, another developing British colony. Pretty soon, Donkin's tin cans would make their way northward.

Going Global

Dwarfed by a massive iceberg, Captain William Parry's wooden ship, the *Hecla*, would still have looked formidable to the four Inuit who approached it from their canoes on 6 September 1820. The

British had presented the group with animal skins and ivory knives that they had packed in the expectation of meeting the inhabitants of the icy wilderness, and the Inuit returned the next day to barter and exchange goods. That evening, when the time came for the crew's dinner, the eldest of the locals watched attentively as a huge tin can of Donkin's preserved meat was brought out and opened by striking the side of the can with an axe and mallet – as was necessary before the first tin openers were invented in the 1840s and 50s. He expressed no desire to eat the meal, 'even when he saw us eating it with good appetites', Captain Parry reported, but after some encouragement took a bite so as not to appear rude. While the mallet was of great interest to the Inuit, 'they did not seem at all to relish' the tinned meat, eating only a little and taking the rest with them on their canoes.[21] Though seemingly not a hit with the local people, these new preserved meats were a valuable resource for Captain Parry's crew, whose voyage into the cold waters to the north of mainland Canada had begun over a year previously when they left England on 4 May 1819.

The British had been looking for a lucrative northern sea-trade route connecting the Atlantic and the Pacific oceans since the sixteenth century, and, after the successful conclusion of the Napoleonic Wars in the early nineteenth century, their search for the North-West Passage intensified. These Arctic voyages required a robust food supply that could last not just for months but for years, as the ships meandered through the ice-packed, island-stubbed seas and narrow waterways of the north. The food they packed aboard their great ships had to sustain the sailors in one of the harshest of environments for human survival. It was not unusual to spend the winter, which in these far northern latitudes lasted for at least half the year, stuck in the ice aboard ship, waiting for the weather to warm up enough to melt the ice and allow sea travel to resume. Captain Parry's crew had been frozen in at a place Parry called Winter Harbour on Melville Island for ten gruelling months. The vegetable element of their diet was provided by sauerkraut and pickles that they had on board, and Parry grew cress and mustard from seed in his cabin, in small boxes 'placed along the stove-pipe'.[22] Food was precious and when fresh meat was found none of the animal was wasted. Occasionally hunting parties found and slaughtered deer: 'it was our custom to consider the heads and hearts of the deer as the lawful perquisites of those who killed them, which regulation served to increase their keenness in hunting, while it gave the people thus employed rather a larger share of fresh meat than those who remained on board'.[23] Captain Parry also recorded the killing of a musk ox on 9 August 1820, all of which – including the flesh of its heart – tasted and smelt 'very strong of musk', but was still a marked improvement on the stores of salted meat. Animal parts that weren't eaten had non-culinary uses. A walrus caught the previous summer provided meat as well as excess blubber which was stored in casks to

Figure 11. A musk ox depicted in Captain Parry's *Journal of a Voyage for the Discovery of a North-West Passage* (1821). The large animal is native to the Arctic, and provided a source of fresh meat for the expedition.

make lamp oil over the winter.[24] For the most part, however, the crew relied on the rations that they'd packed aboard their ships when they left England: salted meat, dry biscuits or flour to make bread, liquor or sometimes beer made of malt and hops, and cans of Donkin's preserved meats and vegetable soups.

From the seventeenth century up until this point, ships' provisions had been relatively standard, relying on only a few types of food that could last long voyages without going bad. In 1677 Samuel Pepys, the famous diarist who was also once secretary to the Admiralty, codified rules on the food that would make up sailors' rations in the navy: 1 pound of biscuit, 1 gallon of (weak) beer each day; 2 pounds of salt beef or salt pork for Sundays, Mondays, Tuesdays and Thursdays, with a pint of peas; three eighths of a 24-inch cod (salted) on Wednesdays, Fridays and Saturdays, with 2 ounces of butter, and 2½–4 ounces of hard Cheshire or Suffolk cheese.[25]* Of these, ship's biscuit is a strong contender for the most commonly complained about food in the early modern period. This biscuit was not the sweet treat that we might imagine, but a simple baked dough, devoid of any fat, moisture, sugar, eggs or yeast so as to keep it from spoiling. Unsurprisingly, it had little flavour, and was so hard it really needed to be soaked in some kind of liquid before being eaten. As American Founding Father Benjamin Franklin simply put it in 1785, 'the ship biscuit is too hard for some sets of teeth'.[26] At sea, eighteenth-century sailors might soak it in a salt-beef stew, perhaps with the addition of onions and seasoned with pepper, which was known as 'lob-scouse' (its popularity in the port of Liverpool perhaps accounts for the nickname 'scouse' or 'scouser' for the city's inhabitants).[27]

---

* In some circumstances ship's biscuit could be replaced with rusk (a hard bread), beer could be replaced by rum or wine, and olive oil given in lieu of butter and cheese. Slight adjustments in quantity were made to the standard provisions in a 1731 document, and the fish provision was replaced with oatmeal. Privy Council, *Regulations and Instructions Relating to His Majesty's Service at Sea* (London, 1731), p. 60.

The biscuit could also be crushed over a 'sea pie' made of meat and vegetables. Similar hard biscuits were supplied to the British army in the Napoleonic Wars, and right through the world wars of the twentieth century, when the notoriously hard biscuits were sometimes carved into touching mementos that reminded the servicemen of home, rather than being eaten (see Figure 12).

Figure 12. Private John William Aucott made a hardtack biscuit into a photo frame and sent it to his mother as a Christmas present in 1916. These hard unpalatable biscuits were a common feature of army and navy rations from the early modern period up into the twentieth century. Their lack of moisture means that many made into souvenirs during the First World War still survive today.

Salted meat, too, had long been a lamented and monotonous feature of the seafarer's diet. Stored in huge casks of brine, when the meat emerged it would, unsurprisingly, hit the nostrils with a 'strong and violent impression of salt'.[28] Sailors would drag it from the ship in nets or barrels through the water, in an attempt to remove some of the salty taste before boiling it up for dinner.

The need to provide more nourishing and tasty food for sailors had inspired culinary innovation for centuries. A kind of portable soup could be made, for example, by boiling down meat in broth until it took on a jelly-like consistency, which was then dried by repeatedly rotating it on a piece of cloth. Cut into chunks and kept dry, as Mrs Elizabeth Dubois claimed in an advert for her 'strong GRAVY SOUP' in 1747, it 'never spoils' and 'is dissolved in a few Minutes in boiling Water'.[29] Just as Appert can't take full credit for the invention of canning, Dubois didn't exactly 'invent' instant portable soup. At the end of the sixteenth century Sir Hugh Plat described a 'victuall for warr' made from a broth of beef, and flavoured with spices, which was boiled down until it became a jelly, and further dried, perhaps with the addition of isinglass* to strengthen it.[30†] A 'pocket soup' appeared also in a 1694 recipe book by Lady Ann Blencowe, who lived in Marston St Lawrence in Northamptonshire, and who described it as being 'of extraordinary service to such as travel in wild and open Countries, where few or no Provisions are to be met with'.[31] Dubois, however, was the first to market her product. Together with an apothecary named William Cookworthy, she was awarded a contract to supply the Royal Navy in 1756. In essence, this instant soup was a precursor of our stock cube, which was invented in the second half of the nineteenth century and sold on a commercial scale in the 1910s. This basic eighteenth-century recipe, called here 'Veal Glue, or Cake Soop', gives an idea of how it was made:

> Take a Leg of Veal, strip it of the Skin and the Fat, then take all the muscular or fleshy Parts from the Bones; boil this Flesh gently in such a Quantity of Water, and so long a Time, 'till the

---

* A type of gelatine derived from fish.

† It would be remiss not to mention that Plat also recommended 'macaroni' (then meaning pasta in general) be served to the seafarers of the time on account of its cheapness and ability to last a long time, claiming that he'd supplied it to Sir Francis Drake and Sir John Hawkins on their latest voyage. Sir Hugh Plat, *Certaine Philosophical Preparations of Foode and Beverage for Sea-men, in Their Long Voyage* (London, 1607).

Liquor will make a strong Jelly when it is cold: This you may try by taking out a small Spoonful now and then, and letting it cool. Here it is to be supposed, that tho' it will jelly presently in small Quantities, yet all the Juice of the meat may not be extracted; however, when you find it very strong, strain the Liquor thro' a Sieve, and let it settle; then provide a large Stew-pan, with Water and some China Cups, or glazed earthen Ware; fill these Cups with Jelly taken clear from the Settling, and set them in a Stew-pan of Water, and let the Water boil gently 'till the Jelly becomes as thick as Glue; after which, let them stand to cool, and then turn out the Glue upon a Piece of new Flannel, which will draw out the Moisture; turn them once in six or eight Hours, and put them upon a fresh Flannel, and so continue to do 'till they are quite dry, and keep it in a dry warm Place: This will harden so much, that it will be stiff and hard as Glue in a little Time, and may be carried in the Pocket without Inconvenience. You are to use this by boiling about a Pint of Water, and pouring it upon a Piece of the Glue or Cake, of the Bigness of a small Walnut, and stirring it with a Spoon 'till the Cake dissolves, which will make very strong good Broth.[32]*

It was the invention of canning, however, that did the most to improve seafarers' diets. Most notably, as Captain William Parry described, tinned provisions lessened the likelihood of their developing scurvy, caused by a lack of vitamin C in the diet. Scurvy had haunted every voyage of the early modern age, killing more than two million sailors between Columbus' 'discovery' of the Americas in 1492 and the mid-nineteenth century. During the Seven Years' War with France (1756–63), 1,512 British sailors were killed in action, but as many as 133,708 died of disease, most

---

* This appears to be the original source for Hannah Glasse's 'pocket soop' in *The Art of Cookery, Made Plain and Easy* (London, 1747), pp. 127–8.

of whom likely fell victim to scurvy.[33] The disease first made itself known with a general, debilitating sense of lethargy and as it took hold gums began to putrefy and blacken, teeth loosened and fell out, bruises appeared as if from nowhere and old wounds reopened. Meanwhile, the sufferer was afflicted by hallucinations and personality changes as their mental state disintegrated. Seizures were common. When scurvy inevitably attacked Parry's ship in that long harsh winter of 1820, the patient, a Mr Scallon, was removed from his rotting bed – which had become sodden with condensation as the crew tried to keep warm below deck – and fed canned vegetable soup, pickles, some preserved currants and gooseberries that they had on board for this purpose, as well as extra lemon juice. Another scurvy patient was admonished after he was found to have been drinking 'pork slush', the salty and fatty dregs in which salt meat was stored, which was the special privilege of the cook, often stored to be sold on to grease dealers on shore for future profit. Parry had prohibited the crew from eating the slushy leftovers on account of his belief (common at the time) that too much salt caused scurvy, and it was seen as more suitable for greasing the masts or burning lamps than for consumption.[34] Although physicians did not understand that vitamin C deficiency caused scurvy until at least the end of the nineteenth century,* several European writers at the time recognised the anti-scorbutic effects of vegetables and fruit (though they also thought that fresh meat, dry conditions and more exercise were cures for the condition). James Lind's experiments in 1747 found citrus fruit to be an anti-scorbutic, and another Scottish physician, Gilbert Blane, is credited with convincing the

---

* The term 'vitamine' was not coined until 1912, as scientists began to depart from the prevailing theory that there were only four essential elements for nutrition: proteins, carbohydrates, fats and minerals. It would take subsequent decades for researchers to work out what these vitamins were and how they worked; in 1933 Norman Haworth described the chemical structure of vitamin C and its synthesis.

Royal Navy to introduce lemon juice to sailors' daily rations in 1795. Sauerkraut, fermented cabbage which is rich in vitamin C, was carried on Captain James Cook's 1768 mission to the South Pacific, and Parry likewise took sauerkraut as well as pickles and vinegar on board the *Hecla* on account of their presumed anti-scorbutic properties. Interestingly, when Nicholas Appert first wrote of his invention, he imagined that tinned food would be used to revive mariners who came down with an illness at sea, 'more especially the worst of them all, the scurvy'.[35]

Captain Parry's ships returned from their voyage safely in November 1820. In the remaining food stores was a 4-pound tin of roast veal. Left uneaten, it ended up in a London military museum where in 1938 scientists finally opened it. By then over 100 years old, the veal was reportedly still perfectly preserved, and was not wasted, but enjoyed by the laboratory cat. Further, ten rats were fed on its contents 'for some days', supposedly providing 'a testimonial to its quality by putting on weight during the period'.[36] The remarkable preserving power of the tin can is also demonstrated in another story from Parry's adventures. The captain went on to make two further voyages in search of the North-West Passage. On his third attempt, in 1824–5, one of the two ships, the *Fury*, became trapped by formidable icebergs in Prince Regent Inlet west of Baffin Island. The ship was badly damaged by the pressure of the ice – so much so that it was deemed to be beyond repair, and the crew transferred to the *Hecla* for the return journey to England. With the *Hecla* already bursting over capacity, some of the *Fury*'s provisions had to be left on the shores near the wreck site in an area that would soon be dubbed Fury Beach.[37] Four years later, in 1829, another Arctic explorer searching for the North-West Passage, Sir John Ross, discovered some of Donkin's tin cans among the abandoned stores from the *Fury*. Reportedly still 'in a perfectly fresh and nutritious state' after years in extreme weather conditions, the cans would help Ross and his crew survive their own Arctic misadventure in which their ship, the *Victory*, became trapped in the masterless ice-water passageways. The group were finally rescued by a passing ship over four years after they had sailed from England.[38] When the company first came across the wreck of the *Fury*, Ross described 'an occurrence not less novel than interesting', finding 'in this abandoned region of solitude and ice, and rocks, a ready market where we could supply all our wants', including supplies of flour, cocoa and sugar alongside preserved meats and vegetables, soups, peas, pickles and lemon juice.[39]

If, in more effectively preserving food, the tin can helped the British to explore and map more of the world, the British also made use of the tin can to transport food from across the world to their own little island in the North Sea without it spoiling. When Appert first revealed his invention, the judging council of a special committee wrote that it would surely 'multiply the enjoyments of the Indian, the Mexican, and the African', and in turn would allow 'an infinity of substances' from across the Empire to be transported to French plates.[40] In a similar vein, the Bristol-born physician Andrew Wynter reflected half a century later that if the tin can were taken up more widely, there would be nothing 'to prevent all the world from pouring its abundance into the lap of England, and her children from becoming the best-fed population on the earth'. Turtle soup, still reserved for the wealthiest 'aldermen and millionaires', could, he reasoned, be made cheaply available from street sellers if sold in tin cans: 'Why should countless turtles lie squandered about on the sands in Honduras?' And, 'Why should not the surplus of salmon of Sweden and Nova Scotia be preserved?'[41] The preserving powers of the tin can were celebrated as a way of accessing, and making use of, the world's resources. As the British Empire expanded, it seemed possible to envisage a global food system, in which surplus foods across the world could be harvested and enjoyed by the British and the other European imperial powers.

Stench and Scandal

Before the stench had time to reach their noses, the Board of Admiralty, arriving to inspect the navy's arsenal at Portsmouth on 14 August 1851, were met with a mob of irate church and state leaders who complained that the smell arising from the meat stores was so bad that it was 'endangering the lives of those who lived in the locality'. As we've seen, the miasma theory of disease still had a grip on the public imagination even at this late date, and the offensive odour prompted fears of a deadly pestilence.[42] The source of the smell was the tin cans made in the factory of the Hungarian-born businessman Stephan Goldner, who had won the contract to supply the British navy in 1845. Two years later, a ration of canned meat became a standard part of sailors' diets. How had these tin cans failed so disastrously in their task of preserving food? And why had the British Admiralty awarded contracts to Goldner in the first place? Such questions were raised with revulsion and great concern in both the nation's newspapers and in the Parliament of 1852, when the case was subject to an official Select Committee inquiry.

How many ill-fated voyages across the now-gargantuan British Empire had failed because their members had been poisoned and starved with corrupted canned meat? Of particular public alarm was the possibility that Sir John Franklin's famous lost expedition had been stocked with Goldner's rotten supplies. Having departed England in May 1845 in another attempt to traverse the North-West Passage, the two ships, *Erebus* and *Terror*, had not been heard of since they were sighted by whalers in Baffin Bay in late July that same year. *Erebus* and *Terror* were huge modern ships, fitted with the latest steam engines that enabled the vessels to get up to 4 knots (roughly 4.6 mph). It was known that Captain Sir James Ross had deposited supplies of tin cans in 1849 at Navy Board Inlet, close to the entry to the North-West Passage, to supply later expeditions to the area. But what if these provisions were also badly preserved? As the Conservative MP Sir William Jolliffe pointed out when petitioning Parliament to investigate the scandal in 1852, if Sir John Franklin and his crew had been supplied with Goldner's unwholesome tins their chances of survival were slim.[43] A number of expeditions had been launched to try to rescue the lost party, including the voyage of the *Assistance* and *Resolute* under Captain Horatio Thomas Austin, but it seems that the canned meat with which they were supplied was untainted. Two years earlier, the grave sites of three of Franklin's crew had been discovered off the east coast of Beechey Island in the Arctic Archipelago. The British continued to hope that Franklin himself was still alive, even promoting him – posthumously, as it turned out – to Rear-Admiral of the Blue in 1852. In 1854 the Scottish explorer John Rae, after speaking with local Inuit hunters, discovered that Franklin's men had perished after both ships had become icebound near King William Island. Rae's leaked report to the Admiralty shocked the nation, especially with its suggestion that many of Franklin's men had resorted in desperation to cannibalism, having access to no better provisions.

Other overseas voyages of the mid-nineteenth century were certainly affected by putrid canned meat. In July 1849 an army unit embarked from Cork for Hong Kong, then a British colony, aboard the *Apollo*, stopping off in Portsmouth to collect its supplies, including Goldner's preserved meats. The soldiers knew it would be a difficult voyage, as they would have to brave four to six months in choppy seas, in a particularly crowded old ship, and in merciless tropical weather. Not long after they set off the ship began to emit an awful pong, which was blamed for an outbreak of malaria, and soon, cholera, too. Emaciated and weak with sickness, the soldiers opened up the tin cans for sustenance, only to find 'the cause of the impure atmosphere' and illness: 'a perfect mass of putridity', which was quickly thrown into the sea. Without enough food on the ship the crew was forced to land at Ilha Grande near Rio de Janeiro, where they regained their strength from fresh provisions, but ultimately 130 men would die of malaria in hospital when they arrived at Hong Kong. British Army officer and politician Colonel Chatterton blamed this sad ending on Goldner's badly made cans.[44]

Government enquiries in the mid-nineteenth century were also concerned about the quality of ordinary salted meat that was supplied to the navy. According to reports, huge amounts of salted beef (and some salted pork) had to be thrown overboard as it was so inedible.[45] In wooden vessels, the ship's stores were certainly not immune to the effects of the constant damp and cold conditions that afflicted the rest of the vessel, and it was not uncommon for spoiled food to be thrown overboard. Sailors routinely opened barrels of salted meat only to find it green and riddled with maggots; cheese went mouldy or simply became so hard as to be inedible, while weevils made light work of turning ship's biscuit into little more than dust. The pursers, who were in charge of managing the ship's accounts, successfully petitioned for an allowance to cover common losses at sea: 'of bread by its breaking and turning to Dust: of butter, by that part next to the Firkin being not fit to be issued: of cheese, by its decaying with Mold and Rottenness, and being eaten with Mites, and other insects: of Peas, Oatmeal and Flower, by their being eaten by Cockroaches, Weavels and other Vermin'.[46] Before the tin can, merchant ships also commonly carried live animals in an attempt to supply the sailors with fresh meat. The experience of the mercantile brig, the *Alacrity*, was typical. As the ship sailed between England and Lisbon in the spring of 1814, most of the livestock on board perished during spells of bad weather, while those animals that survived were so wasted away as to be 'useless', the crew unable to 'recover them from the drenching they had suffered from salt water'. A merchant ship returning from Jamaica in that same year, captained by John Peat, survived on Donkin's provisions, after losing thirteen sheep, five goats, seven kids, ten pigs and fourteen dozen poultry to the stormy seas.[47]

The Goldner scandal reminds us that, even in the mid-nine-teenth century, the tin can was still an emergent technology prone to imperfections. Goldner was variously accused of mixing cheap bits of inedible offcuts of animal offal with the meat, or failing to heat the cans to the correct temperature or for long enough, and of improperly sealing the cans. Later commentators have speculated – probably incorrectly – that Franklin's crew were poisoned by lead that leached from poorly soldered cans. In general, however, canned foods still had a huge advantage over traditional seafarers' provisions as they were less inclined to spoil and provided more nutritional substance. As Britain's global empire grew, so did its reliance on tinned food. By the mid-nineteenth century, several huge canning factories – at Leith in Edinburgh, for instance, and at Bordeaux in western France – were producing tins of soup and vegetables for the use of the respective countries' navies.[48] As Andrew Wynter wrote in 1861, 'England, with regard to her dependencies and foreign countries, is like a city situated in the midst of a desert; vast foodless tracts have to be traversed by her ships, the camels of the ocean; and if these provisions are not entirely to be depended on, the position of the mariners might be likened to the people of a caravan whose water-bags are liable at any moment, without previous warning, to burst, and to discharge the means of preserving life into the thirsting sands.'[49]

Shipping Surplus

As the nineteenth century marched on, technological advances shrunk the world of food even further. Meat processed into tin cans could now travel across the world, feeding sailors and explorers in its furthest corners. But was there a way of transporting *fresh* meat over similar distances, between the countries of the British Empire and its trading partners? The ability of cold temperatures to delay the decay of food had been known for thousands of years. Experiments that attempted to harness this power to move food in ships began in the early part of the nineteenth century and took off in the 1870s. Just as tinned meat had been praised by King George III's household in 1813, chilled meat received the royal seal of approval in 1875 from Queen Victoria. The beef she tried had come all the way from the New York abattoir of Timothy Eastman, who was the first to ship chilled meat from America to Britain. His primitive method involved fanning the cold air from huge slabs of ice, which took up about a quarter of the ship's capacity. The first successful cargo of frozen rather than chilled meat travelled on the French steamer *Paraguay* between Buenos Aires in Argentina and Le Havre in France in 1877. To keep the meat below its freezing point, the ship was fashioned with refrigerating technology in the form of an ammonia-compression machine invented by the French chemist Ferdinand Carré around 1860. In the late 1870s, J. J. Coleman, a Scottish industrial chemist, went into business with the brothers Henry and James Bell who ran a shipping company. In 1880, the steamer *Strathleven* carried aboard a Bell-Coleman machine, the result of their partnership, that used cold dry air rather than ammonia-compression, and wasted none of the 40 tons of frozen beef and mutton that it carried all the way from Australia to England.

Behind these successful missions, of course, were several disastrous, and wasteful, attempts to transport chilled or frozen meats. In 1861 Thomas Sutcliffe Mort established the first freezing works in the world at Darling Harbour in Sydney, which made use of similar technology to that used aboard the *Paraguay*. The Bolton-born emigrant was hopeful that his business would usher in a new era of global food exchange, declaring, 'Yes, gentlemen, I now say that the time has arrived [… when] the over-abundance of one country will make up for the deficiency in another; the superabundance of the year of plenty serving for the scant harvest of its successor; for cold arrests all change.' In Australia and New

Figure 13. An advert for Bell-Coleman Mechanical Refrigeration Co., *c.* 1890. It shows animal carcasses being loaded onto a ship and the novel refrigerating system working within it.

Zealand surplus sheep populations meant that the carcasses were often wastefully made into tallow and, given the low prices created by a saturated market, the offal and pelts were simply thrown away, buried in pits underground. Hoping to open up the international market for Australian sheep meat in England, Mort invested in a project to fit the ship the *Northam* with a new machine in 1876, but to his mortification the vessel didn't even leave the harbour after the pipes that carried coolant leaked ammonia into the meat chamber, destroying all of the food on board. That same year, on 25 December, meat stored in refrigerators made by French inventor Charles Tellier, arriving in Buenos Aires on the *Frigorifique*, made a disappointing festive gift. As it was unloaded, the meat emerged on the deck with 'dark spots' and a 'rather unpleasant' (one imagines, to put it mildly) flavour. Another early refrigerated ship, the *Norfolk*, used technology designed by the Glaswegian-born James Harrison, and set sail from Melbourne in July 1873, but when it arrived at London in October the 20 tons of mutton and beef were found to be entirely unsaleable.[50]

Despite these setbacks, by 1900 nearly half of all the lamb and mutton eaten in Britain was imported from other countries, primarily from New Zealand and Australia.[51] From 1880, Britons might also enjoy butter shipped from New Zealand, on the other side of the world. Surplus beef, meanwhile, was imported all the way from Argentina and the United States. The 350 huge refrigerated ships – 'reefers' – that transported meat and dairy produce across the world at the turn of the century were the first links in the global 'cold chain', which moves food from farm to plate in a series of refrigerators.[52] Huge cold stores were required both at the ports of departure and in the port cities that were the cargoes' ultimate destination. Commercial refrigerators were also being introduced at the turn of the twentieth century in breweries and the meatpacking industry. We learn from Mrs Beeton (with whom we will become properly acquainted in the next chapter) that around 1900 imported beef was available at a lower price than British-grown meat, but with little change in its quality: 'Owing to the great improvement in the means of transport and methods for preserving the meat, carcasses frozen, chilled or refrigerated, arrive in excellent condition.'[53]

With Parliament's repeal of the Corn Laws* in 1846, Britain had embraced the free market – and the opportunity it brought to receive cheaper imported food at the expense of British agriculture. At the end of the nineteenth century refrigerated ships were beginning to bring in fruit as well as animal products. After having discovered the delights of the banana in the

---

* The Corn Laws were tariffs introduced in 1815 to regulate the import of corn in order to keep grain prices high and thereby favour domestic producers.

Canary Islands, London-based grocer Edward Wathen Fyffe began importing Cavendish bananas into Britain in the 1880s. Although they were packaged green in the hopes that they would ripen but not spoil by the time they reached London, it was only the use of refrigerated ships in 1901 that made it possible for the company to confidently ship large amounts of bananas to Britain and Ireland without fearing wastage.[54] With the financial backing of the government, the first ever shipment of bananas in a refrigerated ship took place that year when the Imperial Direct West India Line's passenger liner *Port Morant* arrived in Bristol's Avonmouth docks from Jamaica carrying a hold of large Jamaican bananas. Most Europeans had never heard of the tropical fruit before. Imagine what it must have been like to eat a banana for the first time: bizarrely yellow, creamy, sweet? At the end of the nineteenth century, experiments in bringing over citrus and other fruits grown around Cape Town in South Africa failed as they rotted and wasted aboard. Yet by the 1930s fully refrigerated ships successfully moved fruits from Cape Town, where pre-cooling facilities could hold 6,000 tons, to the port of Southampton.[55] By the mid-twentieth century, a whole host of frozen fruits and vegetables were being imported in reefers to be consumed in Britain out of season. Once again, new methods of food preservation opened up a whole world of food to the British.

———•———

'This pheasant is delicious.'

'I am delighted to hear it,' said the host; 'he gave up the ghost just ten years ago.' 'Nonsense: but this wild duck?'

'Tumbled over with a broken wing, I see by the fracture, in the same year.'

'I suppose,' said a doubting guest, 'you will say next this milk is not foaming fresh from the cow?'

'Milked,' replied our imperturbable host, 'when my little godson was born, that now struts about in breeches.'

'Come, now, what is the most juvenile dish on the table?' was demanded, with a general voice.

'These apples; taste them.'

'I could swear they swung on the branch this morning,' said a sceptic, tasting a slice critically.

'Well, I will give you my word that a flourishing neighbourhood up Paddington way now stands over the field where they were grown.'

'Let us have a look at the water-mark,' said a doubting lawyer, inspecting a canister as he would a forged bill.

There was the date upon it of – what for provisions seemed – a far remote age.

'I shall expect next a fresh olive grown by Horace, to draw on his Sabine wine,' chimed in a poet.

'What a pity we can't bottle up all the surplus brats,' said the father of a family.

'Yes, the day may come when one might order up his grandfather, like a fine old bottle of the vintage of 1790.'

'God forbid!' shuddered the inheritor of an entailed estate.

And so the badinage went on. But we have given enough sterling proof of the value of the intention to excuse a joke or two, and conclude, ere we leave our reader like one of the canisters – an exhausted receiver.

A. Wynter, *Our Social Bees* (1861), pp. 206–7.

Exploration, colonisation and overseas warfare brought with them the need for innovations in food preservation to extend the freshness of natural bounty as it travelled across the globe. The tin can in particular offered European sailors and military men a supply of nourishing meat, untainted with salt to dry it out, as well as scurvy-beating vegetables, that could last the long overseas voyages without spoiling. Preserved beef in tin cans was only officially introduced to the Royal Navy rations in 1847, but this new technology would sustain British soldiers right up until modern times. During the Second World War, when soldiers were without fresh food, they would expect to receive tins of 'bully beef' (tinned corned beef), and 'M & V', a tinned meat and vegetable stew developed by the Aberdeen-based Maconochie Brothers and issued as a military ration during the Second Boer War and the First World War. When conditions allowed it, tins of meat could be heated up with personal stoves called Tommy Cookers which were issued to each soldier. The contents were not always to the soldiers' liking. Bully beef was described as a 'lumpy stew', that 'slopped into one's mess tin', and was paired with those familiar hardtack biscuits, their texture 'like tiles with no taste at all'.[56]

The tin can was not the only innovation in food preservation born in the Industrial Revolution for the benefit of the growing imperial economy. In 1780, the German-born botanist John Graefer became the first to receive a patent for the artificial dehydration of foods, specifically for a 'vegetable of the brassica kind, generally known by the name of green and brown borecole, scotch or other kale'. The vegetables were to be put in a boiling salt-water solution (a pound of salt to 20 gallons of water) for one minute, taken out and hung by small hooks in a heated room until they had dried out. Graefer recommended that the brassica was then put briefly in a damp room so that it wouldn't crumble entirely from lack of moisture. With this method, he wrote, the vegetables would last for at least twelve months and would be especially useful for the Royal Navy for the avoidance of scurvy.[57] Nearly a century later, in 1869, the French chemist Hippolyte Mège-Mouriès invented margarine as a cheaper and longer-lasting substitute for butter. Made from an oil extracted from beef fat, mixed with milk and salt, the new product – known initially as *oleomargarine* – was backed by Emperor Napoleon III who believed it would be a useful durable provision for the French armed forces.* And almost precisely 100 years after this, travel and territorial dominance would again provide the impetus for the development of food preservation technology. When the US-Soviet space race took off in the 1960s, most food given to astronauts was a kind of paste that was squeezed directly into the mouth from a tube. Packets of freeze-dried food were then introduced, which had to be rehydrated by injecting the bag with water. This technology, in which food is frozen and then the water content extracted using low pressure in a vacuum, was adopted and developed for medical purposes during the Second World War, before finding a culinary use in space. It is now a favoured dehydrating method used in the food industry (for example, in instant coffee) as it creates a stable, long-lasting product that preserves its sensorial properties. Next, thermostabilised food that had been heated to destroy pathogens (much like in canning) was sealed in newly created plastic or aluminium foil 'wetpacks' for the first time aboard the *Apollo 8* mission in 1968, when humankind first orbited the Moon. As a treat for Christmas Day the crew were served thermostabilised turkey and gravy, which because the moisture content was preserved didn't even need to be rehydrated (see Plate 11). Technological advances in packaging design also meant that the meal could simply be eaten with a spoon.[58]

Back in the eighteenth and nineteenth centuries, innovation in food preservation not only sustained Britons on overseas missions, but in turn allowed the British and other European powers to exploit the natural resources of far-flung countries by bringing

---

* In the early twentieth century, vegetable oils were made for the first time thanks to the German discovery of hydrogenation, which transforms unsaturated fat to saturated fat. By mid-century margarine was often made with vegetable oils rather than with animal fats, as it is today. Mège-Mouriès went on to sell his patent for margarine to a company who would ultimately, through a series of mergers, become Unilever, now a major multinational and the world's largest producer of margarine.

home a larger and richer variety of food to feed their growing populations. The cold-chain that began to develop in the nine- teenth century with the creation of powerful refrigerated ships is what allows us today to eat imported bananas from Costa Rica, lamb from New Zealand, green beans from Kenya, avocados from Peru, pomegranates from Israel, chicken from the Netherlands and cheese from France. Meanwhile a new focus on agricultural efficiency on British soil was beginning to shape our modern reli- ance on heavy machinery and high-yield crops.

The industrial developments that began in the Georgian era, however, were just the start of the slow transformation of the world's food system. As Wynter's amusing anecdote above implies, the preserving power of the tin can was still a wonder in the Victorian age. And, while refrigerated ships were possible in the 1870s, it took at least fifty years for this technology to become widespread. So how did ordinary people back in Britain expe- rience the world of food, and how did they manage matters of food preservation – and waste – in their day-to-day lives? In the next chapter we're going to explore the streets and kitchens of Victorian London.

4

MRS BEETON AND THE RAG-AND-BONE MAN

Brown soup from tinned mutton.

Ingredients. 2 lbs. of tinned mutton, 2 quarts of boiling water, 1 medium onion sliced, 1 small carrot sliced, ½ a small turnip sliced, a bouquet garni (parsley, thyme, bayleaf), 1 oz. of butter, 1 oz. of flour, salt and pepper.

Method. Turn the meat out of the tin into 1 quart of boiling water, let it remain until quite cold, and remove the fat. Empty the contents of the basin into a stewpan, add another quart of boiling water, boil up, and put in the vegetables, herbs, add a little salt and pepper. Cook gently for 1 hour, and pass the whole through a fine wire sieve. Heat the butter, add the flour, cook gently and stir occasionally until a good brown colour is obtained, then replace the purée and liquor. Boil up, season to taste, and serve. Beef essence, sherry, ketchup and many other things may be added to enrich the soup and improve its flavour.

Time. 2 hours after the fat has been removed.
Average Cost, 1s. 6d. **Sufficient** for 5 or 6 persons.

Isabella Beeton, *Mrs Beeton's Book of Household Management* (1907).

Originally eaten only by sailors, by the end of the Victorian era, tins of preserved food had entered ordinary household larders. Mrs Beeton's famous book included thrifty tips to save food and money for the middle-class housewife.

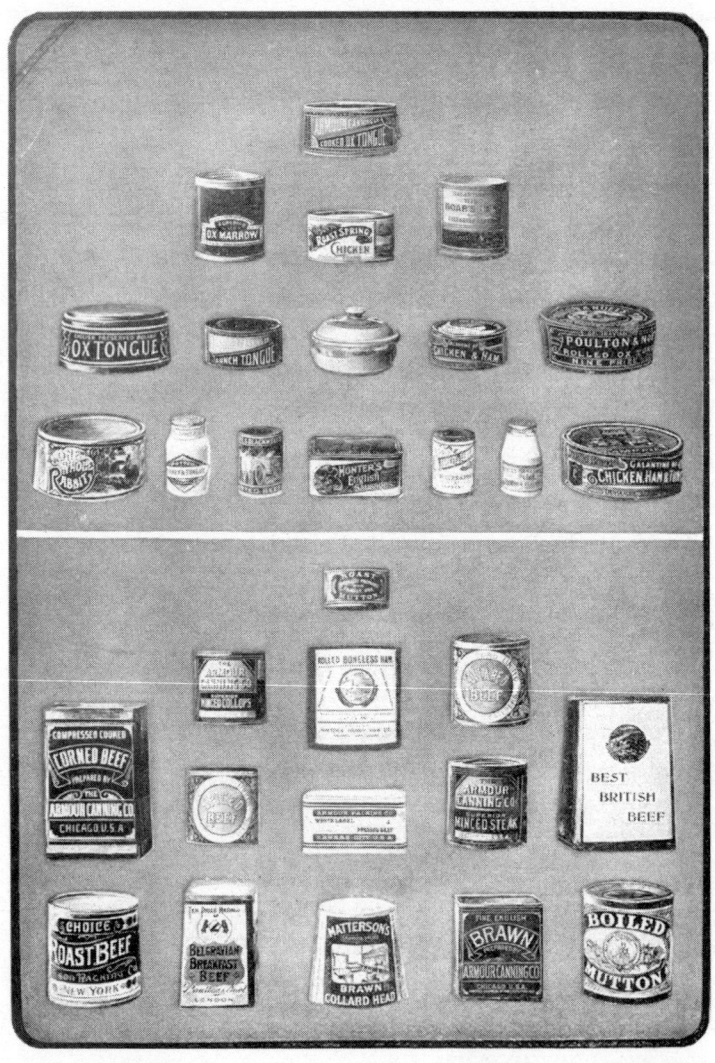

Figure 14. Tinned and preserved meats as depicted in Mrs Beeton's *Book of Household Management* (London, 1907).

———•———

On Sunday morning the London air is crowded with the street-sellers' cries: 'Chestnuts all 'ot, a penny a score'; 'Twopence a pound grapes'; 'Three a penny Yarmouth bloaters'; 'Buy, buy, buy, buy, buy b-u-uy!', yells the butcher. Around a barrow of cabbages, a group of women inspect the produce, as the seller fills his lungs in preparation to shout, 'Where you like, only a penny'. Nostrils are assaulted with the smell of donkey dung, 'chocked with the stench of goats', before being soothed with the sweet 'atmosphere of apples' and the tempting whiff of freshly fried fish. There is a rainbow of fresh food on show to delight the eye: stalls of yellow onions, purple pickling cabbages, white turnips, red meat... A couple of boys run off home for their breakfast, a herring that they've bought for this purpose wrapped up in a piece of paper. Saturday was proper market day, when wages were dished out and the fish sellers still had plentiful supplies of fresh fish. By Sunday, as an ex-fishmonger – a 'keen-looking, tidily-dressed man' – remembers, haddocks that hadn't been sold the evening before were preserved to become 'haddies':

I had a bit of a back-yard to two rooms, one over the other, that I had then, and on a Sunday I set some wet wood a fire, and put it under a great tub. My children used to gut and wash the fish, and I hung them on hooks all round the sides of the tub, and made a bit of a chimney in a corner of the top of the tub, and that way I gave them a jolly good smoking. My wife had a dry fish-stall and sold them, and used to sing out 'Real Scotch haddies', and tell people how they was from Aberdeen; I've often been fit to laugh she did it so clever. I had a way of giving them a yellow colour like the real Scotch, but that's a secret [...]

The market is always busy on Sunday mornings, filled with the noise and bustle of London's working inhabitants who are rushing to purchase their Sunday dinner before the 11 a.m. deadline, when the bell is rung and it's time to put matters of food and flesh on hold for church and a day of rest. Before the stalls have even shut up shop, the ground is strewn with discarded food and the streets are 'green with the refuse leaves of vegetables'.

Scenes such as this were typical of London's inner-city street markets, especially the Brill in Somers Town, an area which, at its busiest, was 'almost impassable' with the crowds and chatter of commerce.[1] The Brill and other markets like it were vividly described by the Victorian journalist Henry Mayhew in his massive mid-nineteenth-century tome, *London Labour and the London Poor*, in which he sought to illuminate the everyday lives of the expanding urban working classes. Fuelled by the Industrial Revolution and the wealth pilfered and extracted from the empire, London at this time was a bursting, bustling, smoke-filled city, alight with industrial manufacturing. The British population had ballooned from 8.7 million at the start of the century to 16.7 million by 1851, and it would reach a colossal 41 million by the end of the Victorian era.[2] For the first time in the country's history, most of these people were living in urban environments rather than the countryside. London's population had more than doubled since 1801; by mid-century it was a metropolis of over 2.3 million that exceeded in size any other city in the world.[3]

These new urbanites were no longer tied to nature's seasonal bouts of plenty and scarcity as they had been for most of human history, and food was no longer exclusively regional.[4] For example, with the construction of steamships and railways, in the middle of the nineteenth century it was increasingly no longer profitable to peddle eggs from neighbouring farmland, as they were brought to central London en mass all the way from France and Belgium.[5] King's Cross Station had opened in 1852, while Euston Station, opening in 1837 just 2 miles away from Smithfield's bustling

livestock market, had become an active cattle handling termi-
nal. Animals that before the industrial age had been marched for
hours from the countryside outside London to Smithfield, were
now sped into the heart of the city before drovers walked them
the last 2 miles to the market. With reforms of sanitation and
urban development, the livestock that became food for Londoners
was increasingly hidden from sight. The sale of live cattle was
pushed outside of the city by 1855, when the live cattle part of
Smithfield's meat market relocated to the suburban location of
Copenhagen Fields in Islington, a site chosen for its proximity
to the newly opened North London Railway cattle station and
the Great Northern Railway depot. What's more, by the 1850s, as
much as three-quarters of the meat sold at the central retail meat
market at Newgate had been slaughtered outside of London and
hauled into the city on huge steam locomotives.[6]

Fuelling the city's reinvention and growth was a mass of
skilled and unskilled workers who had come to the capital to find
work. A rat-catcher who Mayhew interviewed had first travelled
to London from Iver in Buckinghamshire to work on the construction of King's Cross Station. To make ends meet once the job
was completed, he continued to ply his old trade, hunting down
the vermin that scuttled through the city's arteries in search of
morsels of leftover food. The rat-catcher also raised a few bob
from dealing in pickled eels, which he displayed proudly on a
shelf in his room in an overcrowded tenement of Somers Town.[7]

On Monday morning, when the commotion of the weekend's
markets had died down, a new generation of urban middle-class
businessmen arrived in the city's state-of-the-art stations on
packed carriages. These workers – bankers, textile wholesalers,
publishers – commuted in from the neighbouring countryside,
newly created suburbs like Pinner or Surbiton, where many
had moved their families to escape the bad air and dirty slums.
Their wives were the intended audience of Mrs Beeton's *Book of
Household Management*, a tome that rivalled Mayhew's in size

and which was published in 1861, the same year that Mayhew's was published in its final form. The book became essential reading for the newly-wed bride setting up home for the first time, living away from the guidance of family and friends in these new (sub)urban environments. Just like the urban working classes, the town-dwelling middle classes were distanced from the agricultural production of their food. Without the cottage pig, more meat was bought ready prepared from shops than ever before, and inedible leftovers from the domestic kitchen could no longer be dealt with within the household. Mrs Beeton's book offered these new middle classes a guide on how to distinguish themselves from the working population through their meals, manners and household management.

Mrs Beeton, born Isabella Mayson, was herself a Londoner. Her father, Benjamin Mayson, was a vicar's son from the village of Thursby in what was then called Cumberland, who moved to London as a young man to become a wholesaler of cloth, transported from the gigantic warehouses of industrialised Manchester. When he married Elizabeth Jerrom, the couple would bring up their daughter Isabella, the future Mrs Beeton, in his new large business premises on Milk Street in the heart of London's textile district. Following Benjamin's sudden death in 1840, Isabella's mother married the printer and businessman Henry Dorling. In 1857, married herself by then to the publisher Samuel Beeton, Isabella made her debut in print with a recipe for 'a Good Sponge Cake', published in her husband's monthly journal the *Englishwoman's Domestic Magazine*. The recipe was a complete disaster; Isabella forgot to indicate how much flour was needed, and was compelled to write an apology in the next issue, along with the instruction to add flour equivalent to 'the weight of four or five eggs'.[8] Yet, despite her young age (at just twenty-one years old) and her lack of cooking experience, Mrs Beeton's magazine column, 'Cookery, Pickling and Preserving', would become a regular hit under her penmanship, and it was this that inspired the creation of *Mrs Beeton's Book of Household Management*. Containing over 2,000 recipes, along with guidance on how to run a household and comments on the taxonomy, origin and history of animals, vegetables and fruits, Isabella brought together and borrowed advice and recipes from the leading chefs and thinkers of the era.

Although she died at the age of just twenty-eight in 1865 from an infection following the birth of her second surviving child, Mayson Beeton, further editions of Isabella's book continued to spring up with some regularity with additions and edits by Ward, Lock and Taylor, publishers who acquired the copyright to the work shortly after her death. The original forty-four chapters multiplied into seventy-four chapters in the 1907 edition; by this time Mrs Beeton's name had been requisitioned into a brand, a model of good middle-class housewifery. Mrs Beeton remains a household name and copies of the book still reside in many a kitchen, not just in Britain but across the countries of its former empire, passed down perhaps from a grandmother or great-grandmother.

In the mid-nineteenth century London was the wealthiest city in the world, but, as Mrs Beeton and Henry Mayhew's accounts tell us, the riches it amassed from empire and industry were certainly not distributed evenly. So what did food waste and kitchen thrift look like for those in wealth and those in want, living and working in the Big Smoke?

'Specks' and Crumbs

The spiralling population who lived in the cramped city blocks of London often relied on street stalls to provide cheap, tasty and (hopefully) nourishing food. Oysters and hot jellied eels (a dish consisting of eel chunks coated in their own gelatine, commonly flavoured with vinegar) were particularly popular street-food treats for Victorian labourers. Those who made their living selling food from carts were known as costermongers* and Mayhew estimated that there were 30,000 of them – men, women and children – working in London in the mid-nineteenth century. Covent Garden, now known for its boutique fashion stalls, crafted jewellery, artwork and street performers, was once flooded daily with 2,500 costermongers who came to buy wholesale from the famous food market. Just 4 or 5 per cent of the costermongers were classified by Mayhew as 'aristocratic', 'men of superior manners and better dressed' who went around the suburbs selling to higher-ranking people, rather than peddling their wares in the streets. The vast majority, however, were poor.[9]

Among the 'wretchedly poor class' were the street sellers of trotters. By the Victorian era, these and other innards were considered lowly foods. The 1883 issue of the popular publication *The Girl's Own Paper*, aimed at middle-class women, listed condescendingly 'pig's feet' among the foods consumed in poor districts, along with cow heels, tripe, 'black puddings, small savoury pies' and 'cheap fish, including messels, whelks and cockles'. 'We hardly ever see lentils, haricot beans or maize', it lamented, and this despite attempts to make these supposedly more agreeable dishes popular.[10] Most of London's trotter-sellers were elderly women or children, many of them immigrants from Ireland, perhaps dispossessed in the Irish potato famine that had blighted the country between 1845 and 1852. In the mid-nineteenth century, when Mayhew was writing, the street-trade in sheep's trotters was growing. Fifteen or twenty years previously the fellmonger, who dealt in sheepskins, had sold the trotters on to make glue. As glue-making techniques improved and its price decreased, it became more financially lucrative to cook the trotters and sell them on to the poor street-sellers as food instead.

---

* When it was first recorded in the sixteenth century, the word costermonger originally meant 'apple seller', a 'costard' being a type of large cooking apple.

At the so-called 'trotter yard', the feet were scalded first for half an hour, the hooves scooped out of the trotters and sold on to become manure or to the manufacturers of Prussian blue, a bright pigment used by painters. Women would then work to scrape off the hair before leaving the feet to boil for around four hours. The street-sellers sold the finished product around public houses and at the back doors of theatres. According to Mayhew, one trotter-seller lived with 'another old woman' in a single room costing 1 shilling and 4 pence a week, making between 3 and 6 shillings per week in the trade. Her own diet was a cheap one, consisting mainly of fish and bread. One night she sold a sheep's trotter to a woman in a pub who 'put her teeth into it' before declaring that it 'wasn't good', and proceeding to fire off verbal abuse. Luckily, the landlord intervened: 'I'll not have this poor woman insulted; she's here for the convenience of them as requires trotters, and she's a well-conducted woman,' he said. 'Why, who's insulting the old b---h?', fired back the angry customer. 'Why, you are [...] and you ought to pay her for her trotter, or how is she to live?' responded the landlord. When the woman continued to refuse to pay, the landlord instead offered the seller sixpence, and others too came to her rescue, even offering her double the price for her trotters. 'It was the best trotter night I ever had', the old trotter-seller remembered.[11]

Mayhew also wrote of 'Irish refuse sellers' who would collect 'shrivelled, dwarfish, or damaged fruit' that was thrown aside by fruit-salesmen in the markets and sell it in poor neighbour-hoods.[12] Known to the Victorians as 'specks', similarly misshapen or blemished produce is today dubbed 'wonky' fruit and veg by the British media. Under increasing pressure to 'cut endemic food waste', and with the softening of EU laws monitoring aesthetic standards in vegetables and fruits, supermarkets began to resell these rejected foodstuffs as distinct products from 2014.[13] The Victorians, too, made use of unsightly fruit and vegetables – in their case to pay the wages and fill the stomachs of the poor who traversed London's grimy streets. Poor, middle-aged or elderly women, many widowed, paid to the fruit salesmen a quarter of the price for discarded 'specks': 6 pence to a shilling for a bushel of apples, for example, or anything from 1 shilling 6 pence to 2 shillings 6 pence for the more exotic plums. Sometimes the fruit was only slightly damaged – perhaps a little over-ripe or an area of it raided by the burrowing of wasps and other insects. If the street sellers were really lucky the salesman might throw in a few bits of good fruit among the 'refuse'. Mayhew recorded the words of one 'stout, healthy-looking woman' who had worked selling fruit for at least twenty years. 'The childer is my custhomers,' she said, 'such as has only a ha'pinny now and thin, God hilp them.' Young rascals with a taste for misbehaving openly mocked the pope, knowing that the Irish sellers were Roman Catholics. They would 'come a mile' to visit the speck seller, she said, 'for they know it's chape we sill!'. In the Victorian era poor children would have been put to work to help families afford to buy the food they needed to survive. Though the trade in damaged fruit seemed at first 'great an evil' to Mayhew, he reasoned that since specks were only sold to those in poverty and their 'poor children' who, 'were it not for the halfpenny and farthing lots of the refuse-sellers, would doubtlessly never know the taste of such things', it could not be condemned. With an established name for herself, Mayhew's 'stout' woman was able to make 5 or 6 shillings a week from her deals in wonky fruit.[14]

It wasn't just old fruit and vegetables that were sold among London's poor in the busy streets around Petticoat Lane, the centre of a sprawling slum, but crusts of stale bread. A part-time baker's assistant told Mayhew how he began selling old bread to customers, dressed 'some in rags, and some as decent as their bad earnings'll let them'. The women who bought from him did so to feed their children, concerning whom the seller had often heard them say 'they must make haste, as the poor things are hungry, and they couldn't get them any bread sooner'.[15]

Life in the cramped and polluted Victorian slums, especially those clustered in the East End, was indeed notoriously difficult. In *Oliver Twist*, Charles Dickens had described the broken windows of Jacob's Island in Bermondsey, perhaps the most notorious of the London rookeries. The rooms here were 'so small, so filthy, so confined, that the air would seem to be too tainted even for the dirt and squalor which they shelter'.[16] As cholera plagued the capital, Mayhew dubbed the slum 'pest island', a shockingly insanitary place where 'swollen' animal carcasses littered the streets, 'the whole air reeked with the stench of rotting animal and vegetable matter', and the water was dyed 'as red as blood' with the pollutants of the nearby leather tanning industry. The backyards of any house lucky enough to have one were taken over by pig-sties, while roaming chickens and ducks 'fatten[ed] on offal' from muckheaps.[17] It was normal, even for the regularly employed poor, to fill rented rooms with five or more people, and many slum-dwellers had no access to cooking facilities, buying instead what they could from street vendors or cook shops. Unsurprisingly, children of the slums were undernourished and infant mortality rates were high. Cramped housing projects sprang up not just in London but across many of Britain's swelling towns and cities, as a result of the unprecedented rapid population growth of the mid-nineteenth century. As a matter of survival, not a scrap of food could be wasted in these challenging environments, opening up the trade in 'specks' and crumbs to poor costermongers.

Buying and Scavenging

Waste that to modern sensibilities seems entirely useless had an afterlife in Victorian London, making its way into the hands of street-sellers who sold it on to rag-and-bottle shops. Cluttered with piles of rags, wastepaper and books, old clothes, broken saucepans, sacks of bones, tubs of kitchen grease and rows of dusty bottles, these were insalubrious places where little outside light could penetrate. 'The stench in these shops is positively sickening', Mayhew bemoaned. In *Bleak House*, Dickens, a contemporary to Mayhew, described a rag-and-bottle shop situated in the shadow of Lincoln's Inn in Holborn, owned by an old, disorganised drunk named Krook:

> In one part of the window was a picture of a red paper-mill, at which a cart was unloading a quantity of sacks of old rags. In another, was the inscription, BONES BOUGHT. In another, KITCHEN-STUFF BOUGHT [...] In another, WASTE PAPER BOUGHT [...] In all parts of the window were quantities of dirty bottles: blacking bottles, medicine bottles, ginger-beer and soda-water bottles, pickle bottles, wine bottles, ink bottles [...][18]

In poor neighbourhoods, where a rag-and-bottle shop was found on nearly every street, the owners worked to find a home for the scraps they took in, selling some bits to local people and other materials to various tradesmen.

By Mayhew's time, rag-and-bottle shops employed people to buy dripping and other leftover 'kitchen stuff' directly from better-off households. As in the Middle Ages, the leftover dripping from joints of meat as they roasted was often given to the cooks or servants as a kind of perk or addition to their wages, and these employees could then sell the leftovers on to make more money. From at least the seventeenth century, women, characteristically described as wearing large cloaks and carrying baskets to collect their wares, had gone from door-to-door each morning seeking out dripping. 'Any kitchen stuff have you maids,' was their distinctive call. However, the Victorian mistress of the house became increasingly suspicious of these backdoor deals, imagining that they might lead to the pilfering of 'silver spoons', 'candles' and

Figure 15. A scene from inside Krook's rag-and-bottle shop
in Charles Dickens' *Bleak House* (1852–3), showing piles of papers,
rags and signs to the right inscribed with 'Kitchen Stuff'
and 'Bones Bought'.

butter, among other things. Charles Dickens, the son of the famous
author of the same name, wrote in 1879 that the cook might even
be inclined to direct 'surreptitious stabbings' at the household
meat in order to squeeze out more juice that she could sell on for
a profit to the 'kitchen stuff' buyer.[19] The mistress therefore kept
a close eye on the selling of the household staff's 'perquisites', or,
alternatively, the cooks were permitted to take the dripping from

roasts to the rag-and-bottle shop directly in return for a small fee. The dripping was then reused by the poor as a substitute for butter, the latter increasingly preferred by the growing middle classes as a sign of culinary and cultural refinement. Smothered on a piece of bread, dripping instead made a tasty and filling breakfast for London's working classes.

Of a morning, cooks in reputable houses might also hear the cry of the hare-skins buyer: 'Any hareskins, cook? Hareskins.' Traders in hare skins relied on wealthy households, as hare was a food reserved for the elites: 'I never tasted hares' flesh in my life', one buyer noted, though 'I've smelt it when they've been roasting

Figure 16. The buyer of 'kitchin stuff' was a common sight in London from at least the seventeenth century. This is a late-eighteenth-century drawing based on *The Cryes of the City of London*, first published around 1688.

them where I've called.' Instead, the man lived on 'bread and butter and tea, or milk sometimes in hot weather', with perhaps the occasional piece of fried fish. There were seldom enough hare skins available to sustain a hare-skin buyer for a whole year, especially during the summer months, so many of them worked simultaneously flogging other goods like flowers, matches, or pots and pans. When they were able to get their hands on hare skins, they sold the crude product on at a small profit – perhaps half a penny each – to hat-furriers (or their suppliers), who felted them into fashionable top-hats.[20]

The bones left over from consumption were, by the nineteenth century, an increasingly sought-after commodity. The introduction of clover and turnips into British agriculture in the eighteenth century, as part of the Norfolk four-course system, was helping to increase the amount of nourishing nitrogen in the soil. As it slowly decomposed, bonemeal could be converted into another key plant nutrient, phosphorus, which would further increase the productivity of farmland. For this purpose, bones were even imported from mainland Europe, especially from Germany. By the 1840s, the agricultural scientist John Bennet Lawes had worked out that adding sulphuric acid to bones would make superphosphates, turning bonemeal into a faster-acting fertiliser.[21] Bones from butchers and the knacker's yard were collected by men employed by bone-grinding mills. On the street, poor men who focused on collecting bones to sell on to become manure were better called street-foragers than buyers. Mayhew described them as old and infirm. They were known to approach poor children who had collected discarded bones as they played in the streets, bartering toys in exchange for the youngsters' grisly findings.

Kitchen leftovers of all kinds were sought after by hogs' wash buyers. According to Mayhew, these remains included: 'of the scum and lees of all broths and soups; of the washings of cooking utensils, and of the dishes and plates used at dinners and suppers;

of small pieces of meat left on the plates of the diners in taverns, clubs, or cook-shops; of pieces of potato, or any remains of vegetables; of any viands, such as puddings, left in the plates in the same manner; of gristle; of pieces of stale bread, or bread left at table; occasionally of meat kept, whether cooked or uncooked, until "blown", and unfit for consumption'. Our use of the term 'hogwash' today to mean 'nonsense' or 'rubbish' reflects the poor quality of these scraps, which were fit for pig food rather than humans. One of the principal hogs' wash sellers in Victorian London was a pig dealer named Jemmy Divine from Lambeth, who used the remains he collected to feed his pigs as well as selling them on to other pig keepers. Divine made deals with hotel owners and taverns whereby they sold him their kitchen slop at on average 20 pounds a year. He employed workers to do the rounds on a horse and cart equipped with two or three tubs 'well secured, so that they may not be jostled out' into which the hogs' wash could be collected. Many independent street-buyers noted the value in hogs' wash, and went door-to-door collecting it from taverns, hospitals, poor houses and private homes. Rather than watch it go down the drain, modest eating establishments and houses might even give their kitchen detritus away to the buyers, so desperate were they to swiftly cleanse their residences of its stench.[22]

As we've seen, pigs had long been employed to snaffle up household food waste. But prototype systems of household waste removal in London had by the end of the eighteenth century developed into a more organised communal provision using dustmen. The name lingers today, a reference to the 'dust' that was generated from the daily fires lit in this period, and which made up the major part of the household's refuse. Each dustman was employed by a wealthy 'dust-contractor', who bartered for contracts from local parish authorities. Wearing the distinctive 'fantail hat' with flaps to cover their necks and shoulders, dustmen would come through the London streets each morning

DUST O!

Figure 17. The dustman from William Marshall Craig's
The Itinerant Traders of London (London, 1804).

in a dust-cart, traditionally ringing a bell (though this noisy practice became illegal in 1839) and calling 'DUST O!' to alert the community to their presence. Once collected by horse and cart, the refuse was taken to huge 'dust yards' outside of the city.

Whereas today seagulls and smells are about all you'll sense among the mountains of rubbish at our gigantean landfill sites, the Victorian dust yard swarmed with poor women and girls. Sifting through the rubbish, their faces covered with layers of soot and their clothes permeated with its blackness, these scavengers collected anything that could be useful in wicker baskets. The ashes they found could bring in a profit when sold on for manure, the cinders for fuel and bones went to burning-houses to be prepared for agricultural use.[23] Bits of leftover meat or vegetables could be sold on as manure or animal feed, rags could be transformed into paper as well as manure for the hop fields, while any fat they found might earn the workers a small fee from the soap boilers or glue makers. Chickens and pigs also rooted and pecked at the garbage heaps, hoping to find edible scraps. One contemporary commentator observed that the women and children 'sifters' were reduced to a 'disgusting and degrading' life not much better than the dust-yard swine, while another shuddered at 'the sight of such wretched squalid specimens of humanity and womanhood'.[24] Quite often the 'sifters' were the wives and children of the dustmen themselves, and were paid a small wage by a subcontractor, dubbed a 'hill-man' or a 'hill-woman'.[25]

Perhaps the most wretched of all the groups Mayhew described were those who scavenged in the streets for scraps of food, both to sell on and to consume themselves. 'Many of the very old live on the hard dirty crusts they pick up out of the roads in the course of their rounds, washing them and steeping them in water before they eat them', he wrote. Other street 'finders' hoped to come across a morsel of leftover bread thrown away by the servants of a wealthy home, or that a friendly housekeeper might offer them some bones from which they could scrape off and consume a few flecks of remaining flesh. Lack of food, Mayhew reasoned, had led the minds and energy of these unfortunate people to decline just as their bodies withered from malnutrition. The rag-and-bone man, clad in a ragged coat and broken shoes, would carry a 'greasy bag' on his back, usually with a stick in hand which had a spike at the end to sift through rubbish and pick up bits of useful waste that could be flogged at the rag-and-bottle shops. Realistically 5 lbs of rags or bones would get him 2 pence, and he could not hope for much more than this from a day's work. Rag-and-bone men would walk 20–30 miles in a day, aiming for enough useful finds to supply them with food for the day and a

Figure 18. Mayhew included this illustration of the 'sifters' searching in the large metropolitan dust yards for sellable leftovers.

Figure 19. Mayhew's rag-and-bone man, based on a photograph by Richard Beard.

temporary shelter for the night. As well as bones and rags, these homeless men were even reduced to seeking out dog excrement or 'pure', so called because it was used in the leather industry to purify skins on account of its alkaline nature. The street finders could get themselves as much as 8 to 10 pence per bucket at the tanyards clustered around Bermondsey and its slums.[26]

Middle-Class Thrift

The poverty that plagued the street buyers, sellers and scavengers who we have met so far was seen by much of Victorian society as the self-imposed result of careless financial decision-making. 'Much ignorance [...] with respect to all household knowledge' was rife among the poor, according to Mrs Beeton, who recommended that middle-class housewives visit the homes of those in poverty in order to instruct them – in a 'pleasant and unobtrusive manner' – in good household management and cookery.[27] To Mrs Beeton's middle-class readers, a superior knowledge of culinary thrift set them apart from those in poverty, for 'frugality and economy are home virtues, without which no household can prosper'.[28] Here, clearly, the term 'thrift' had transitioned from its original sense of 'thriving' to refer to the frugal practices that the middle classes associated with respectability.[29] Ever since the celebrated chef Alexis Soyer had established soup kitchens in Bethnal Green and Spitalfields, feeding the poor had become an important expression of middle-class morality that acted also as an opportunity to impart culinary knowledge. Soup was the food of choice as it offered a lesson in how to turn wholesome simple ingredients, or foods that would otherwise be wasted, into nourishing meals.[30] By a law of 1834, the poor were legally divided into the 'deserving', those who were physically incapable of working through no fault of their own, and the 'undeserving', those who could physically work but were deemed – by virtue of not having work – too lazy or ignorant to do so. The same attitude underlay the policy of sending the most destitute to the workhouse as a way of making them earn their food and learn the value of hard work. Conditions in the workhouse were deliberately appalling to deter as many people as possible from accepting such charity. Despite his sympathy towards London's poor and the work he did to expose the wretchedness of their living conditions, Mayhew was writing for a middle-class audience who liked to see themselves as superior in habits and morals to the urban working populace. In an era when change was so rapid, the hierarchy of social class was something that the middle classes believed could be transcended and climbed, if only with the right determination. 'Simple industry and thrift', declared Samuel Smiles in his book *Self-Help*, published in 1859, will 'go far towards making any person of ordinary working faculty comparatively independent in his means.'[31] Many of Mrs Beeton's readers, meanwhile, were middle-class women who themselves hoped to climb the social ranks through thrift to become ladies of leisure, removed from the relentless rounds of kitchen and household work, which in the wealthiest homes was carried out by a whole team of paid staff.

The new generation of industrialised middle-class housewives who read Mrs Beeton's *Book of Household Management* could certainly not afford to waste food, and a well-managed household was essential to make the most of the money and food that passed through its doors. 'Great care should be taken that nothing is thrown away, or suffered to be wasted in the kitchen, which might, by proper management, be turned to a good account', the book stated.[32] Famously, Mrs Beeton compared the housewife to the 'commander of an army', overseeing a hierarchy of house-hold workers: in the female roles were the housekeeper, who was designated as 'second in command' to the mistress, a cook (a woman in all but the wealthiest households, where male chefs were employed), kitchen maids who assisted the cook, scullery maids who scrubbed floors and prepared poultry for the cook, and a housemaid. The book includes a table of suggested wages for each worker, and the mistress or housekeeper is advised to keep an account book to keep an eye on any wasted expenditure. The housekeeper would also keep expensive commodities like sugar and spices under lock and key to monitor their use and to prevent pilfering. This was an aspirational book, however. Many of Mrs Beeton's lower-middle-class readers – including perhaps Isabella herself in the early years – would have had no cook and no housekeeper, but only a couple of maids (perhaps a maid-of-all-work and a nursemaid for the children). Much of the household and kitchen toil would still fall, therefore, on the shoulders of the mistress.[33]

Reducing food waste was to begin long before the food reached the table. The mistress herself, or preferably the housekeeper, who oversaw the work of the cook in the kitchen, was instructed to save money and food by choosing joints of meat of which the butcher had a superfluity.[34] Especially in the 'sultry' summer months, without the cooling relief of an electric refrigerator, it could be difficult to find meat that had not begun to turn. It was vital to learn how to recognise tainted product, especially for those new urbanites who were cut off from fresh countryside produce. It was, Mrs Beeton noted, imperative that the housekeeper or mistress examine meat that came to the house: 'if flies have touched it, the part must be cut off, and the remainder well washed'.[35]*

In the home, each item of food was to be stored in a clean and tidy place where it would keep fresh for as long as possible to avoid it going to waste. Vegetables kept best on the stone floor of the larder†, and rice and seeds 'for puddings' should be covered

---

* In the 1907 version, 'well washed' was replaced with the more specific instruction that the meat be 'well wiped with a clean cloth dipped in warm water and vinegar'.

† The larder gets its name from 'lard', a reference to the importance of lard in preservation. Allen, *Can It!*, p. 41.

in cloth to prevent insect infestation (though 'even this will not prevent them from being affected by these destroyers, if they are long and carelessly kept').[36] To prevent cheeses from drying out, they were to be wrapped in a damp cloth, and put in a pan with a cover in a cool storage area or larder.[37] Fresh butter was to be wrapped in white paper, or the housewife could make use of the new 'red brick' butter dishes, into the wide lid of which cold water was poured and changed regularly in the hot summer in order to keep the butter cool.[38] Other cookery book writers of the time had similar food-storing ideas: nets suspended in the store room could house lemons and oranges, while onions, shallots, garlic and various dried herbs should be hung up with ropes from the ceiling for winter use.[39] According to S. Beaty-Pownall, who published a series of cookbooks at the turn of the twentieth century based on a previously popular magazine series, preservatives like jams, pickles and chutneys should be stored on the lower shelves of the larder where it was cooler, so as to improve their shelf life.[40] Meat was hung in the larder on hooks to soften the flesh. Knowing when to retrieve it was a skill learned through experience – by smelling and touching: 'Though it is advisable that animal food should be hung up in the open air till its fibres have lost some degree of their toughness,' Mrs Beeton wrote, 'if it is kept till it loses its natural sweetness, its flavour has become deteriorated, and, as a wholesome comestible, it has lost many of its qualities conducive to health.' Once the flesh showed the 'slightest trace of putrescence', it was time to cook it, but meat that had started to taint could supposedly be saved by washing it in vinegar or another type of disinfectant like a solution of potash/potassium permanganate.[41]

Mrs Beeton was writing for the first generation of middle-class women who were more likely to buy their meat ready smoked from the butcher than preserve it themselves in the laborious ways of their Tudor forebears.[42] Nevertheless, her guidebook is bursting with instructions on how to preserve meat and other foods in traditional ways that were still practised in the countryside. By now London was a smoggy, filthy city; the devastating Broad Street cholera epidemic of 1854 and the accompanying Great Stink of 1858 were still fresh in people's minds. Distanced from the source of what they ate, the industrialised Victorians were also understandably anxious about the adulteration of food as it moved through the supply chain. The chemist Frederick Accum's *Treatise on Adulteration of Food and Culinary Poisons* of 1820 first ignited the public's anxiety about the substances hiding in London's food, and a hugely thorough report from the journal the *Lancet*, published each week between 1851 and 1854, confirmed the true and shocking extent of adulteration. Alum was added to bread to add bulk and make it appear whiter, milk contained chalk (again to whiten it), artificial colouring for Gloucester cheese might be tainted with red lead, and sulphate of copper could be found in some pickles and bottled preserves on account of its pleasing colour. Some of these substances were poisonous and even deadly. Just a year before Mrs Beeton's *Book of Household Management* was first published, in 1860, the first Adulteration of Food Act marked an initial attempt to regulate such practices, but it was by all accounts a failure, given that it relied on members of the public to pay a fee so that suspected provisions could be tested, which they were only on the whim of the local authority.[43] Only in 1875 was an Act passed that made it universally illegal to adulterate food with harmful chemicals or substances other than those expected to be in the product, and it took until 1925 for the law to specifically target harmful chemical preservatives.[44]

No wonder, then, that Mrs Beeton's book trumpeted the virtues of a return to a supposedly simpler and healthier rural lifestyle. Bacon, then as now, was a particularly popular preserved meat that Mrs Beeton recommended be made in a traditional way. It was first salted in a 'cool' and 'well ventilated' place and covered with 'bran, or with some fine sawdust'. Hung over the fire to smoke, this formed a protective layer against the heat of the fire that burned for a month or so below the flitch. Wiltshire bacon had by this time a reputation 'as the finest in the kingdom'. To imitate it, the mistress could sprinkle each flitch of bacon with salt, letting the blood drain from it for twenty-four hours. Next, 1½ lb of coarse sugar, 1½ lb of bay-salt, 6 oz of saltpetre* and 1 lb of common salt mixed together was rubbed in to the flesh of the whole pig, which was turned every day for a month, before being hung up to dry and smoked for ten days.[45†]

For the Victorian housewife, according to Mrs Beeton, the preservation of fruits for winter began in spring with oranges, which could be made into orange wine. Summer fruits like gooseberries, currants, strawberries and raspberries could then to be transformed into jams and jellies.[46] Other fruits were preserved in syrups or spirits. Pickles, preserves and other sauces were to be labelled in stoneware, earthenware, or later, glass jars, and neatly stored on a separate shelf in the cool larder. Nothing, in fact, showed the difference between a 'tidy thrifty housewife and a lady to whom these desirable epithets may not honestly be applied' more than 'the appearance of their respective store-closets'. With an ordered larder, the good housewife would easily be able to reach for the desired bottle at a moment's notice, so that no time was lost and 'no dish spoiled for the want of "just a little something"'.[47] Mrs Beeton even included a drawing to demonstrate how the jars might be labelled. This one (Figure 20), for an 'India Pickle' marked 'PICCALILLY', a word that arrived in the UK in the mid-eighteenth century, denotes a British take on a South Asian relish of pickled vegetables and spices, evidence of Britain's now far-reaching empire.

---

* Saltpetre (potassium nitrate) only appears in bacon curing methods from the eighteenth century, sometimes in fact in harmful volumes. As previously noted, before this, bacon was commonly dry salted or preserved in brine.
† With the advent of refrigeration, the Harris company came up with a brine which preserved the meat faster. This wet cure is what remains famous today rather than the original dry salting method.

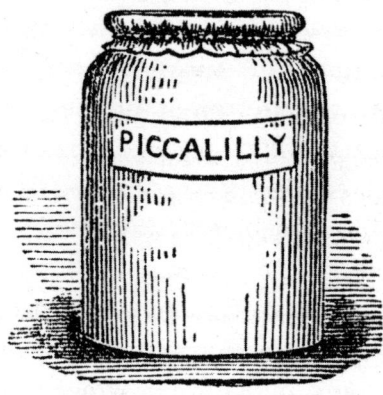

Figure 20. An example of Mrs Beeton's labelled jars to store pickles
and preserves, for an 'India pickle' called 'piccalilly'.

Many preserves, like fruit jams and pickled vegetables, had
been staples made in the still-room since Tudor times to siphon
off some of the summer's glut for the winter table, but imperial
expansion brought new examples and flavours to British palates.
Ketchup, for instance, arrived from south-east Asia where it was
created as a way of using up pickle juice in which fish or vegetables
had been preserved. Variously described in Britain as 'catsup',
'catchup' or 'kitchup', it likely took its name from the Chinese
word 'kôe-chiap' meaning a brine of pickled fish. Probably a thin,
salty and vinegary brown sauce, it was first encountered by British
travellers and colonisers at the end of the sixteenth century and
soon imitated at home. It wasn't until the end of the nineteenth
century that ketchup more commonly referenced a type of
thick tomato sauce that we might recognise today. Before this,
Victorian home cooks adapted the original recipe, creating a thin
sauce based on mushrooms, walnuts, oysters, anchovies, or even
lemons or elderberries, with the addition of spices.[48] Mrs Beeton
appears to have been particularly fond of the popular mushroom
version, which in her recipe was made by layering mushrooms

and salt in a pan alternately (to each peck* of mushrooms, ½ lb of salt) and leaving them for a few hours before breaking them up. For three days the mushrooms were left, occasionally being stirred or mashed so as to extract their juices. Next, cayenne, allspice, ginger and pounded mace at the ratio of 1/4 oz, 1/2 oz, 1/2 oz and two blades respectively for every quart of the extracted mushroom-liquor was added. The mixture was then transferred to a stone jar and placed in a saucepan of boiling water for three hours, then in a second stewpan for half an hour. It was cooled in a jug for a day, poured into another jug and strained into clean bottles. Finally, a few drops of brandy were to be added for each pint of ketchup.[49] The sauce could then be kept for a long time, and used in a variety of dishes to add flavour, making it perfect for Mrs Beeton's labelled larder jars.

Chutney was another hugely popular condiment, a thick sauce made from fruit or vegetables and spices, which was adapted from original Indian recipes in the British Raj with available local ingredients. Using vinegar rather than the fermenting power of the hot Indian sun, British chutneys allowed cooks and housewives to use up surplus produce and preserve it for future use. In essence, Mrs Beeton's 'Mango Chetney' called for the mixing and heating of these ingredients: 1½ lbs of moist sugar, ¾ lb of salt, ¼ lb of garlic, ¼ lb of onions, ¾ lb of powdered ginger, ¼ lb of dried chilies, ¾ lb of mustard-seed, ¾ lb of stoned raisins, two bottles of best vinegar and thirty large unripe sour apples.[50]

Much of the appeal of Mrs Beeton's book lay in the thrifty recipes and advice it imparted. When it came to cooking, Mrs Beeton recommended frugal dishes by including the word 'economical' in brackets. She also offered the reader a choice between a standard version of the dish and a cheaper one. Moreover, in recipes labelled 'cold meat cookery', Mrs Beeton showed how 'leftovers' or 'remains' of meat could be used up the following

---

* Equal to 2 imperial gallons.

day. When the cook of a middle-class home created the daily menu for the mistresses' approval, it was important that she make use of any leftovers from the day before as well as considering any ripe, perishable produce from the garden. Without a microwave, still a century away, any remains of 'jugged hare' (hare stew) from the night before should be rewarmed by putting it in a covered jar in a saucepan of boiling water: 'this method prevents a great deal of waste', Mrs Beeton stated.[51] She recommended that any leftover cold roast or boiled beef was potted, to be consumed at the breakfast table, where it would join a broad selection of other cold meats and hot dishes, from mutton chops to sausages and eggs. The method for potting meat was much the same as it had been for centuries: meat was pounded in a mortar with a little bit of butter, seasoned with ground cayenne and mace, and then put into a glass or earthenware pot with clarified butter poured on top to create a seal.[52] The leftovers from roast beef also made good fritters:

BEEF FRITTERS (COLD MEAT COOKERY)

INGREDIENTS. The remains of cold roast beef, pepper and salt to taste, ¾ lb. of flour, ½ pint of water, 2 oz. of butter, the whites of 2 eggs.

Mode. Mix very smoothly, and by degrees, the flour with the above proportion of water; stir in 2 oz. of butter, which must be melted, but not oiled, and, just before it is to be used, add the whites of two well-whisked eggs. Should the batter be too thick, more water must be added. Pare down the cold beef into thin shreds, season with pepper and salt, and mix it with the batter. Drop a small quantity at a time into a pan of boiling lard, and fry from 7 to 10 minutes, according to the size. When done on one side, turn and brown them on the other. Let them dry for a minute or two before the fire, and serve on a folded napkin. A small quantity of finely-minced onions, mixed with the batter, is an improvement.

Time. From 7 to 10 minutes.

Average cost, exclusive of the meat, 6d.

Seasonable at any time.[53]

Cold roast beef could also be 'hashed', as in a recipe that called for one teaspoonful of tomato sauce, one teaspoonful of Harvey's sauce (a bottled sauce based on vinegar, soy sauce and ketchup), '1 teaspoonful of good mushroom ketchup, ½ glass of port wine or strong ale, pepper and salt to taste, a little flour to thicken', and '1 onion finely minced' to be added to a stewpan 'with whatever gravy may have been saved from the meat the day it was roasted'. Simmered for ten minutes, the fat was to be skimmed off the top before the beef was coated in flour and laid in the gravy mixture to be warmed for five minutes. 'Serve very hot, and garnish with sippets* of toasted bread', Mrs Beeton concluded.[54] Fritters, hashed meat, or croquettes were similar dishes that were popular among the middle classes, who could afford the meat from which they were made but who needed still to employ some culinary thrift. 'Rissoles', moreover, reused cold roast beef by covering it in breadcrumbs, frying it to 'a rich brown' colour and serving it with gravy.[55] Bones from meat were also redeployed. The shank-bones from mutton, notably 'so little esteemed in general', were useful in giving 'richness to soups or gravies, if well soaked and brushed before they are added to the boiling', while 'roast-beef bones, or shank-bones of ham, make excellent stock for pea-soup'.[56]

Similar thrifty recipes continued for the section on fish dishes. Leftover cod broken into flakes worked 'very well' for a dish of

---

* Small pieces of bread used for dipping in a sauce.

'Cod a La Creme', for instance. One ounce of butter, one chopped shallot and a little minced parsley were added to a quarter of a teacup of white stock, which was thickened with a little flour and added then to quarter of a pint of milk or cream. With a dash of cayenne, powdered sugar and lemon juice the sauce for the fish was complete.[57] Desserts too could be fashioned from leftovers. What do you do with the egg yolks if you're making a recipe that requires egg whites alone? Mrs Beeton instructed the reader to use them to make a 'pudding or a custard'. Served with custard, an old stale cake could be remade into a 'tipsy cake', by soaking it in 'sweet wine or sherry' to moisten it. Alternatively, stale cake could be crumbled, soaked and 'heaped over' with whipped cream, so that it became a trifle.[58] Finally, the cheese course: old, dry cheese that was pounded in a mortar with fresh butter, then potted with a coating of clarified butter, would last several days more. As many thrifty cooks still do today, Mrs Beeton recommended using up the hard cheese at the rind in 'Welsh rare-bits' (cheese toasties) or melted as a sauce to cover macaroni.[59]

We can see how food was preserved and recycled in the middle-class kitchen in these two sample menus for family dinners served in January:

Sunday. 1. Boiled turbot and oyster sauce, potatoes. 2. Roast leg or griskin[*] of pork, apple sauce, brocoli, potatoes. 3. Cabinet pudding,[†] and a damson tart **made with preserved damsons.**

Monday. 1. **The remains of turbot** warmed in oyster sauce, potatoes. 2. **Cold pork**, stewed steak. 3. Open jam tart, which should have been made with the **pieces of paste left from the damson tart**; baked arrowroot pudding.[‡][60]

---

[*] The lean part of a pork loin.
[†] A type of steamed bread and butter pudding.
[‡] Bold type is my own emphasis.

New Thrift, New Waste

In the first edition of her *Book of Household Management* of 1861, Mrs Beeton noted how the use of a 'refrigerator' in the larder could help keep freshly picked fruit fresh until you were ready to turn it into a traditional preserve.[61] In the 1907 version, this one reference to the refrigerator had grown to ten. As Mrs Beeton's lengthy description makes clear, this was not an electric refrigerator as we would know it, but rather an 'ice box', a small wooden cupboard for keeping a few items of food cool, which Britons had begun to import from America in the 1840s. There, ice was harvested from huge frozen lakes using cutting tools and horse-drawn carriages before being shipped across the ocean insulated with sawdust (not without, of course, a percentage of loss in the form of melted ice). Norway was another key exporter of ice to British households. Early iceboxes were made with hard wood and looked much like a cabinet or chest of drawers. The wood walls were lined with zinc, insulated with asbestos, felt or slag wool, and a separate section was kept topped up with blocks of ice, with a drain to take away the water as the ice melted to another chamber beneath the refrigerator. This drip tray, unsurprisingly, had to be regularly emptied. Mrs Beeton recommended thriftily making extra use of the waste water container as an ice bath to keep bottles of liquor cool.[62] To ensure a well-stocked icebox, ice was delivered by the 'ice man', who went from door-to-door on his horse-drawn cart.

In 1834, the first ever refrigerating machine was displayed publicly by the American inventor Jacob Perkins in London's Fleet Street. Rather than relying on a continual supply of ice blocks, his invention created artificial ice in a vapour-compression cycle using liquid ammonia. Hundreds more patents for refrigerators would be filed in the next century as inventors took up the challenge of artificial cold-making. As I've already elucidated, huge refrigerated steamships were traversing the oceans by the end of the Victorian era with refrigerated beef, mutton, dairy and bananas on board. London's meat markets also depended on refrigeration as by 1890 most of the city's meat came in to the city frozen rather than being slaughtered on site.[63] However, even after the invention of the first domestic artificial refrigerator in 1913, it would still be at least half a century until the majority of British people had artificial refrigerators in their kitchens (a revolutionary development we shall explore in Chapter 6). Still, Mrs Beeton's later edition shows us that in the early 1900s the middle classes were becoming more reliant on the preserving power of cold. And as the icebox became more popular over the century, the huge amounts of ice that needed to be supplied spurred on further developments in the creation of artificial refrigeration.

Figure 21. An insulated wooden ice box, known as the dry air siphon refrigerator, which cooled food in the late Victorian home via blocks of ice inserted into a separate compartment. Made by the Seeger Refrigerator Co., St Paul, Minnesota, USA, 1880–1920.

Another monumental advance in food preservation, the tin can – whose development and early use we analysed in Chapter 3 – was also changing household diets by the end of the Victorian era. As steamships connected Britain with food markets on the other side of the world, ordinary people ate more and more meat, much of it from the surplus stocks bred in Australia and New Zealand, and much of it tinned. In the 1861 edition of her book, Mrs Beeton warned that meat preserved in tins was 'sometimes carelessly prepared' and even adulterated, leading to the poisoning of those 'poor men' of the navy who were its chief consumers. The 1907 edition still cautioned that 'The Nutritive Value of Tinned Meat is less than that of fresh meat, and it is somewhat insipid' in flavour, but nevertheless included a whole chapter on its use to the housewife. 'Tinned meats, soups, fish, poultry, fruit and veg-etables now occupy an important place in our food supply', Mrs Beeton's imagined voice acknowledged, and they provide 'handy substitutes when fresh provisions are difficult to procure', either being out of season or otherwise in short supply. In this later edition of the book, recipes that called for tinned meats include ox-tail soup, white soup (made from a tin of rabbit), curry (using tinned lobster), and 'beef roll' (from tinned roast beef, to which chopped bacon, herbs and egg were added before the whole thing was shaped into a roll and baked). A supply of tinned sardines should always be kept in the store cupboard since they could be whipped up into a variety of dishes, Mrs Beeton reasoned from the page.[64] Tinned calf's head was the key ingredient in a quick-fix version of a quintessential Victorian dish, mock turtle soup. Live green turtles were imported at great cost from the West Indies in huge numbers to fulfil a demand for turtle soup that had begun in the late eighteenth century, when the dish was popular at lavish formal occasions. Unable to afford this luxury, middle-class Victorians could recreate its flavour by using calf's head, the flesh of which mirrored the texture of turtle meat. Still with its skin on, the calf's head was scalded and the brain removed, before being boiled for an hour wrapped in a cloth to release the meat from the bones. Along with butter, ham, onions, shallots, mushrooms, herbs, around a pint of stock and a dash of wine or sherry, the meat was simmered for two hours, and the resulting soup was served with forcemeat balls.[65] A simpler option, in the later edition of Mrs Beeton's famous work, used instead half a tin of calf's head:

MOCK TURTLE SOUP

Ingredients. ½ a tin of calf's head, 2 ozs. of ham, cut into dice, 1 medium-sized onion sliced, 1 small carrot sliced, 1 or 2 strips of celery, a bouquet garni (parsley, thyme, bay-leaf), 1½ ozs. of butter, 1½ ozs. of flour, sherry, lemon-juice, forcemeat balls, [...] salt and pepper, 5 pints of boiling stock or water.

Method. Melt the butter in a large stewpan, fry the ham and vegetables until lightly browned, and sprinkle in the flour. Let the ingredients cook slowly until well browned, and meanwhile drain the calf's head, add the liquor to the stock or water, and cut the meat into neat pieces. Pour the boiling stock or water over the browned vegetables, boil up, skim well, and, when the vegetables are tender, pass the whole through a fine sieve or tammy*. Replace in the stewpan, bring to the boil, season, add sherry and lemon-juice to taste, put in the prepared meat and forcemeat balls, and serve when thoroughly hot.

Time. From 1½ to 2 hours. **Average Cost,** 2s. 6d., exclusive of the sherry. **Sufficient** for 6 or 7 persons.[66]

Tins were to be stored in a cool place, and when ready to be opened they were cut along one side (presumably with an early can opener, which were essentially 'lever knives' with a hooked

---

* A cloth used as a strainer.

blade to pierce the tin) and a small hole made in the other so that the contents could more easily be pushed out. Tinned meat could be heated in the tin in boiling water, although Mrs Beeton advised that the meat be served with sauces so as to disguise the off-putting tinned taste.[67] Canned vegetables, especially peas and beans, were met with less suspicion and regularly used. By the 1907 edition of Mrs Beeton's book, tinned tomatoes and fruits, like pineapple chunks and peaches, were already recognisable store-cupboard ingredients, as they are today.

Figure 22. Images of tinned meat and tinned fruit from Mrs Beeton's *Book of Household Management* (1907), by which time they had become store-cupboard staples for ordinary people.

Concerns like those expressed in Mrs Beeton's tract, about the adulteration or changed nature of food preserved in tin cans, fuelled the development of chemical preservatives at the end of the nineteenth century. Unlike the tin can, chemicals could

preserve food without altering its look and with these new preserving chemicals there was no need for meats and butter to be
heavily salted. By 1900 a government report stated that 39 per
cent of foods contained chemical preservatives, or 'antiseptics'
as they had come to be known (the compounds were also used in
medicine for treating wounds). Boric acid was the most common
chemical preservative that was added to a whole host of perishable foods including dairy and meats, while salicylic acid (from
willow-bark) increased the shelf life of jams, wine and non-
alcoholic drinks, and formalin was sometimes found in milk and
meat products.[68] The 1907 edition of Mrs Beeton's book suggested
that the housewife herself could add a little boric acid to cream
and milk as a preservative that worked by 'neutralizing the lactic
acid'.[69] We now know the substance causes abdominal pain,
vomiting and diarrhoea, and, worse, created the conditions in
which the deadly disease bovine TB could thrive.* Otto Hehner, a
leading figure in the Society of Public Analysts, was among those
doctors and other experts at the time to call for a ban on the use
of chemical preservatives, of which, in 1900, he said 'I do not see
why a grocer should drug me'. In time many came to view these
compounds as a new form of adulteration, rather than a solution
to the problem, and they were slowly regulated over the course
of the first half of the twentieth century.[70] A novel 1925 law laid
out which types of foods and drinks could contain preservatives,
limiting these to sulphur dioxide and benzoic acid, and in what
quantities. Further, sausages, pickles and sauces, coffee-extract,

---

* Today, boric acid is an approved additive (E284) but in the narrow context
of preserving sturgeon caviar. It is also used in a type of food packaging as
a liquid absorber to extend shelf-life. Annalaura Lopez et al., 'Evolution of
Food Safety Features and Volatile Profile in White Sturgeon Caviar Treated
with Different Formulations of Salt and Preservatives During a Long-Term
Storage Time', Foods, 10 (2021), p. 850; EFSA, 'Safety Assessment of the
Active Substances Carboxymethylcellulose, Acetylated Distarch Phosphate,
Bentonite, Boric Acid and Aluminium Sulfate, for Use in Active Food
Contact Materials', EFSA Journal, 16(2), (2018), p. 5121.

and wines which were permitted to contain these chemicals had to display this information in a label, much as is the case today.[71]

The adulteration of foods was in part what led to the boom in branded products at the end of the nineteenth century, as customers looked for names they could trust to sell 'pure' or 'unadulterated' foods. Dairies, like London's Welford and Sons Co., advertised their milk as 'absolutely pure & free from preservatives', for example (see also, Plate 14).[72] Still a well-known brand today, Cadbury's used the controversy to sell more products, launching 'unadulterated Cocoa Essence' in the late 1860s, and comparing it with competitors' brands like Fry's, whom they accused of adulterating products with additives including sago and sugar.[73] As Britain moved into the Edwardian era, householders continued to replace home-made produce with ready-made mass-produced versions; the role of the still-room waned as pickles and preserves could now be bought off the shelf at the grocer's. Although Mrs Beeton looked back wistfully on the wholesome traditional preservation techniques of the country housewife, even in her original first edition she was already reaching for the packaged shortcuts of the industrial age. She noted, for instance, that pickles could be purchased from shops, perhaps for an even cheaper price than the home-made versions, but that it was still better to attempt the latter if possible.[74] A similar thing was happening to ketchup. By the mid-nineteenth century tomato versions (often also containing anchovies) had largely replaced the mushroom or fishy varieties, and in 1876 from Pittsburgh, Pennsylvania an American company launched Heinz Tomato Ketchup, bringing the bottled form to the British housewife in 1886. Soon, cans of the now quintessentially British Heinz Baked Beans were being made in the UK, first at their Harlesden factory from 1928. By 1888, too, a thick meat extract made from Argentine beef had been introduced into British homes in its soon-to-be iconic red-labelled jar. Bovril was the creation of John Lawson Johnston, a Scottish entrepreneur working in Canada, who supplied the French army with a million cans of 'Johnston's Fluid Beef' during the 1870s as a novel way of preserving and transporting meat. In the Victorian household Bovril would became a quick flavouring for sauces, gravies and stews. The working classes were used to making 'beef tea' from meat scraps and gravy, but by adding water Bovril became instant beef tea, removing the need to make it at home with leftover ingredients. Other industrial beef extracts had been made (often from animal hides that were otherwise wasted at tanners and tallow-makers in Australia) after the German scientist Justus von Liebig recommended its health benefits, founding a company that would later invent Oxo cubes. Bovril claimed to be superior to other beef teas and meat extracts (see Plate 13), containing added dried meat powder that was rich in nourishing protein. Mrs Beeton's 1907 cookbook included more ready-made condiments such as Bengal mango chutney (costing 1 shilling per bottle), bottles of Harvey's sauce, Worcestershire sauce, mushroom catsup and essence of anchovies.[75]

All these new products and brands were helped to market with the invention of new machinery at the turn of the century, which made it easier and cheaper to produce the bottles they were sold in.[76] By 1890, the folding paper carton had also arrived in Britain, providing a new way to brand and protect mass-produced foodstuffs. Perhaps unsurprisingly, with these new and exciting products came more packaging waste. Consumers were encouraged by manufacturers to keep buying more branded products, and jars, bottles, cartons and food scraps that would have been reused in the earlier part of the Victorian era more often went to waste. The housewife was encouraged to liberate herself from the time-consuming tasks of preserving and recycling food scraps, and by the end of the Victorian period more and more people were throwing away the containers in which they received their pre-made condiments, rather than reusing them.[77]

In London, where many, in order to live, struggle to extract a meal from the possession of an article which seems utterly worthless, nothing must be wasted. Many a thing which in a country town is kicked by the penniless out of their path even, or examined and left as meet only for the scavenger's cart, will in London be snatched up as a prize; it is money's worth.

Henry Mayhew, *London Labour and the London Poor,* II, p. 8.

As the Victorian metropolis grew, leftovers were fed back into the growth of the city. Ash from kitchen fires and discarded oyster shells went to make new bricks to build more houses, while animal and vegetable food waste became manure and fertiliser to grow food for more maws. As Mayhew's journalistic banquet has shown us, many of London's poor found employment in trades that reused or recycled kitchen waste, reselling otherwise discarded food to the working classes, or selling on refuse to be reimagined and repurposed in industry. In the middle-class home, too, Mrs Beeton peddled what we in the modern age would call a 'circular economy'* of food in which leftovers are seen as a resource to be reused rather than as something to be thrown away. The leftovers from fancy joints were restyled at the breakfast table in the type of household Mrs Beeton was writing about, while any surplus was preserved to be consumed at a later date. Unwanted or unused kitchen stuff from richer homes made its way into the hands of the poor where it took on a useful after-life. Away from the self-sufficient agricultural communities of previous generations, when food was cyclically produced and then reused as crop and livestock-feed, Victorian Londoners still established ways of recycling the outputs of the food industry in a uniquely urban way.

---

* This language was first used in the 1980s and is the opposite of the 'linear economy' in which food is simply produced, consumed and thrown away.

By the end of the nineteenth century, culinary innovations, like the icebox and branded condiments, had seeped into middle-class homes, and wages were growing across the social spectrum. Changes in how people lived and how they ate took place at a faster rate than they had in any previous period of history. With more wealth came more meat-eating, for example, more tinned food, as well as a more varied diet that adopted flavours and ingredients from across the British Empire, fast approaching its peak size of covering a quarter of the world's landmass. Waste expanded as affluence did, to the point that rag-and-bottle shops began to close, finding little value in the food scraps and packaging waste that had once sustained the urban salvage market. Instead, refuse collection was increasingly managed by local authorities. In 1875 the Public Health Act instructed households to place their refuse in a dustbin and gave local authorities the responsibility of removing it, though this was technically still on a voluntary basis. Cards given out to help people interpret the new regulations in their area suggested that the 'house refuse' that contractors were obliged to remove included 'cinder ashes, potato peelings, cabbage leaves, and kitchen refuse generally'.[78] In London, the Public Health Act was enforced in 1891.

Late Victorian authorities were puzzled by what to do with the growing mountains of rubbish. Towns were struggling to find space for it in tips, and with the new public health initiatives, these fly-infested dust yards were increasingly condemned as vectors of disease. Victorian writers looked with mounting concern on the work of the dust-sifters who we met earlier in this chapter, women and children who slaved away on rotting heaps of waste, seeking the pickings from 'the filth of those more fortunate'. In 1901, the engineer W. F. Goodrich wondered 'how such a filthy and degrading process could have any advocates at all in the light of our modern sanitary science'.[79] Rather than reusing or recycling household waste, late Victorian authorities preferred to incinerate it. The brainchild of the aptly named Alfred Fryer, this technology was invented in 1874 in the form of a machine equally aptly named 'the destructor'. Fryer worked in a Manchester sugar refinery, and had invented a similar machine to incinerate spent sugar cane, before realising that the technology could have a wider use. Incineration was seen as a more sanitary and modern alternative to leaving the refuse to be picked at and to decay in tips where it would become a further hazard from the perspective of public health. It was also an efficient way of destroying diseased animal carcasses or rotting food.[80] Later in the century, new types of incinerator used the steam emissions that were created in the process of incinerating waste to generate electricity, but with mixed success since they often required more fuel to reach the high temperatures needed to work effectively.[81] Of course, destructors also left behind their own pollutants, discharging fumes and a solid waste called clinker. As a result, they did not entirely replace rubbish heaps in towns.[*] In the Victorian countryside, where rubbish collection was far more ad hoc, and since plenty of land was available, tipping continued to be the normal way of disposing of solid waste. Whereas, as we saw in Chapter 2, late Victorian public health reforms clamped down on urban pig-keeping in the cramped backyards of the slums, country pigs were still employed to chomp down household food waste.

Traditional practices continued on a smaller scale into the twentieth century. In some districts, you might even occasionally still see a rag-and-bone man looking for scrap today. Nevertheless, by the Edwardian era the waste from London kitchens was no longer routinely funnelled through the recycling-based street economy, a pattern mirrored in microcosm

---

[*] Today, municipal waste incineration is also criticised on account of its waste emissions, in this case greenhouse gases and especially carbon dioxide.

across the country's smaller cities and towns. But this frivolity would be short lived. The Great War – the 'war to end all wars' – was brewing on the Continent. With conflict came scarcity and a new era in the history of food waste and kitchen leftovers.

5

THE KITCHEN FRONT

POTATO PIGLETS

6 medium well-scrubbed potatoes
6 skinned sausages
Cooked cabbage – lightly chopped

Method Remove a centre core, using an apple corer, from the length of each potato, and stuff the cavity with sausage meat. Bake in the usual way and arrange the piglets on a bed of cooked cabbage. (The potato removed from each is useful for soup.)

Ministry of Food, *Potato Pete's Recipe Book* (1943).

During the Second World War, the Ministry of Food encouraged the consumption of potato, an energy-rich food that could be grown at home, and which was useful for bulking out meals, replacing fat and flour in pastries and for masking unpleasant flavours. *Potato Pete's Recipe Book* – complete with an anthropomorphised potato character wearing a green hat and chewing on a piece of straw – showed the housewife a number of ingenious food-saving recipes (see Plate 15). In this example, the potato core was reused in a soup, and even the water in which potatoes were boiled was saved for this purpose.

— • —

On the eve of the First World War, Britain had developed not only an advanced internal railway network, but an international system of free trade that brought in food from across the world at prices that local farmers could no longer compete with. Strikingly, only around 3 per cent of British farmland was used to produce grain for flour. With cereals making up a large part of an ordinary person's diet, 58 per cent of the population's calories came from produce imported from overseas, as did 81 per cent of Britain's bread wheat.[1] Domestic dairy production had also dramatically declined over the last part of the nineteenth century as cheap foreign imports of butter and cheese flooded the market from Europe and as far as New Zealand. Though the Germans certainly produced more of their own grain and bread, food imports accounted for roughly half of the calories consumed there also, three-quarters of which came into the country by ship.[2] Early on in the war, Britain preyed on the weakness of this new international system of trade, using the Royal Navy to prevent ships from reaching German ports in what become known as the 'hunger blockade'. This tactic did its work to limit Germany's food supply, leading to widespread food shortages and hunger. The blockade was even enforced for months after Germany surrendered, and used to put pressure on the Germans to accept the severe terms of the Treaty of Versailles. As a 'fact-finding commission', organised by Winston Churchill, then serving as secretary of state for air and war, ruled in February 1919, as long as 'Germany is still an enemy country, it would be inadvisable to remove the menace of starvation by a too sudden and abundant supply of foodstuffs. This menace is a powerful lever for negotiation at an important moment.'[3]

But Germany, too, would use food supplies as a weapon of war. German U-boats retaliated by targeting British ships from 1916, with naval propaganda stating the rationale: 'England is about to starve [...] in the near future England will lie on the ground: unconscious, hungry, beaten with the same weapon with which it attempted to defeat the dutiful German people.'[4] In February 1917, the U-boat campaign destroyed nearly 500,000 tons of merchant shipping bound for Britain; severe food shortages became a real possibility.[5] In response, the British government sought to limit imports of non-essential foodstuffs and to become as self-sufficient as possible by transforming any available soil into fertile cropland and encouraging people to grow their own food.

In this time of scarcity, any food that people did have was not to be wasted; for those left on the Home Front, the prudent management of their eating habits was seen as a vital contribution to the war effort. A proclamation from King George V urged all people to 'practise the greatest economy and frugality in the use of every species of grain'. Households were exhorted to reduce their consumption of bread by at least a quarter, and to avoid using flour for pastry or anything other than bread.[6] As one poster from the First World War directed the housewife, 'DON'T WASTE BREAD! SAVE TWO SLICES

Figure 23. 'DON'T WASTE BREAD!' poster made by the Ministry of Food, Clarke and Sherwell, Ltd Printers, London, *c.* 1917.

EVERY DAY and Defeat the "U" Boat'. In another poster, an imaginary 'Mr Slice O'Bread' lamented that each day the 'bit left over' and the 'waste crust' was absent-mindedly discarded by Britons. 'If you collected me and my companions for a whole week you would find that we amounted to 9,380 tonnes of good bread – WASTED!' By wasting or consuming bread when it wasn't needed, the reader was 'adding 20 submarines to the German Navy!' Mr Slice O'Bread declared.[7]

By the time war came again in 1939, Britain had reverted to its global system of trade, so that 70 per cent of its food was imported across the oceans in merchant ships.[8*] The Allies cut off Germany and Occupied Europe from the world's food supplies once again in August 1940, but the wartime collapse of the global food system brought shortages of foodstuffs for Britain as well. No bacon, cheese and butter could be imported from Denmark, no onions from Spain and France. Since valuable cargo space could not be wasted on fruit imports, oranges became a thing of the past in Britain; the Crown colony of Cyprus, which would otherwise send its produce to Britain, was left with an excess of the fruit. Meanwhile, in Australia, a self-governing 'dominion' within the British Empire since 1907, a surplus of apples and pears which were normally destined for international trade was left to rot on trees.[9] Once again, those on Britain's Home Front were expected to live with less food, and to waste as little as possible. Though starvation never blighted the British population, bellies certainly rumbled. Past culinary delights, imported from the ever-shrinking world, were suddenly out of reach. Writing

---

* This compares to 46 per cent today based on 2020 statistics. Defra, 'United Kingdom Food Security Report 2021', 22 December 2021, https://www.gov.uk/government/statistics/united-kingdom-food-security-report-2021/united-kingdom-food-security-report-2021-theme-2-uk-food-supply-sources#:~:text=Headline-,In%202020%2C%20the%20UK%20imported%2046%25%20of%20the%20food%20it,%C2%A321.4%20billion%20was%20exported.

The Lion, the Witch and the Wardrobe in 1939, C. S. Lewis remembered sugary Turkish delights – 'sweet and light to the very centre' – so fondly that he imagined Edmund would give up his family for bites of their exotic flavour.[10] Back in the real world, food shortages were at their most severe over the cold winter of 1940–41. Yet again, the message to those on the Home Front was 'waste not, want not'. The Ministry of Food commissioned a series of posters with snappy maxims and directions. 'BETTER POT-LUCK with Churchill today, THAN HUMBLE PIE under Hitler tomorrow: DON'T WASTE FOOD!', declared one, the incensed face of Hitler on the pie dish made recognisable by his characteristic moustache. By 1940 it was a criminal offence to wilfully waste food or drink that could have been consumed by people. In one instance, the two managers at a bakery in Swinton, Lancashire were prosecuted for feeding 11 tons of bread to greyhounds.[11] Global combat meant a war on waste would have to be fought on the Home Front.

Rationing

Not long after the First World War began, in 1914, Britain's high streets filled with queues as people sought out enough food to feed their families. Panic buying and hoarding soon set in. Initially, the government was reluctant to interfere with the public's liberty in making dietary choices, but in 1916 they deemed it necessary to introduce some limits on consumption, restricting lunches in public eateries outside of home to two courses and dinners to three.[12] That year, the Ministry of Food was set up to regulate the nation's food supply, and Lord Devonport was named minister of food control. By February 1917, as the impact of Germany's unrestricted submarine warfare began to be felt in Britain, Devonport instituted voluntary rationing in an attempt to reduce the waste of existing resources. Each person was asked to limit their consumption of bread to 4 pounds per week, with 2½ pounds of meat and 12 ounces of sugar. Door-to-door canvassing encouraged householders to take a 'food pledge', which went: 'In honour bound, we adopt the national scale of voluntary rations.'[13] This soft tactic proved ineffective, however, against looming shortages; long queues soon led to complaints and riots. Edgar Waite, who lived in Sunderland during the First World War, described how, since there was as yet no enforced rationing,

> it was very difficult getting hold of food, especially meat. And women had to queue up very early in the morning. Somebody would say, 'Now, there's a butcher's shop up the road there; they've got some meat.' And they would queue up hours before the butcher's shop opened, on the off chance of perhaps only getting a bone with a bit of meat on. They had to just accept anything that's going.

Since there was 'no ration scheme in operation', Edgar concluded, 'it made things very difficult to purchase either cigarettes, beer or food'.[14] Dorothy Ann Haigh remembered leaving her family home in Bournemouth early in the morning to queue up for meat, taking her little brother – one of ten siblings – to stand in a butcher's queue while she waited in another, hoping that at least one of them would manage to get something to take back to their mother to cook up. As in many towns, tensions sometimes rose to uncomfortable heights. 'We was all queued up for margarine outside the Maypole in Southampton', Dorothy continued, when a group of men 'forced the doors open', demanding that the women be served.[15] On a larger scale, George Hodgkinson remembered a mass protest in Coventry, when the parading crowds carried banners declaring, 'Equal Food, Equal Distribution All?'. He explained that 'this was a veiled threat': 'There was not proper distribution of food, unless there was a proper regard at top levels for the need for reasonable allocations and so on.'[16] Underlying such unrest was a sense that the little food that was available was being unfairly distributed and wasted. When strikes began in Manchester, sweeping across the engineering sector, workers had read of 'bacon lying rotting at the Port of London, or herring in the north of Scotland, or of potatoes being in some places superabundant, and in others, non-existent'. These tales, unsurprisingly, bred 'deep resentment' among those who suddenly found it a struggle to get hold of enough food for their usual dinners.[17]

Compulsory rationing was finally introduced in London in early 1918 under a new food controller, Lord Rhondda, and the scheme was extended across the nation by the summer. Each person was to register at their selected shops (in an era before supermarkets this meant the butcher, fishmonger, greengrocer and baker) in order to allow the shopkeepers to receive as much food as necessary for their customers' rations. There was some variation in the weekly civilian ration, but it tended to include 8 oz of bacon (rising to 16 oz in the summer of 1918), 5 oz (about half a packet) of margarine or butter, 8 oz of sugar, 4 oz of jam and 2 oz of tea, while fresh meat was rationed by price.[18] Food-wasting behaviour, including throwing rice at weddings or feeding stray dogs, was made illegal.

With the food shortages of the First World War still within memory, the government was quicker to impose legal controls on people's diets in the Second World War. Food rationing was introduced in January 1940 with limits on butter, bacon and sugar, and by August 1942 almost all foods other than bread and game meat like rabbit and pigeon were rationed. Vegetables and fruits were also not included on the ration, but many – like bananas, lemons and oranges, which needed to be imported from faraway warmer climates – were extremely hard to come by. Other foods, such as tinned goods, biscuits and cereals, were rationed using a points system, each person being allocated sixteen points per month (later increased to twenty) to spend on them. The number of points each item cost changed depending on national stocks. In 1942 a tin of sardines might cost you seven points, for example, while baked beans came in cheaper at four points a can.[19] Moreover, even though bread itself was not put on the ration until the war had ended, by 1942 the only type you could buy was the 'national loaf', an unappetising salty (given the preservative properties of salt) mush of wholemeal flour to which calcium and other vitamins were added. The ration allowed a fairer distribution of food across the population, and was particularly beneficial to the working classes who could now access a balanced diet that included a mix of vitamins, calcium and protein.

But eating from the ration was commonly challenging for those on the Home Front. Mrs Lettice Miller was a mother of three young children during the Second World War. Rationing, especially of meat, butter and sugar, was hard for the family, she remembers, and the diet 'dull and monotonous'. Households in this era had cold larders, and perhaps a cold food cabinet, but without freezers and with limits on the amount of sugar everyone was allowed, preserving foods to make them last longer was difficult, Lettice remembers. Yet because the family lived in the countryside they had access to more fruits and vegetables than town and city dwellers. Lettice could bottle fruit to preserve it over the winter. 'I used to spend a frightful lot of time picking fruit, and I used to pick pounds and pounds of blackberries, which we [...] used to bottle,' she recalled, while she made as much jam as possible on the limited sugar ration. The children gratefully received an extra allowance of orange juice and cod liver oil, and at one point some blackcurrant purée.[20]

Living in London, Vere Hodgson recounted in her diary on 9 July 1940 the news that butter and margarine were to be limited to just 6 oz per person per week, along with 2 oz of cooking fat. 'Ugh!', she wrote, frustrated. She recalled how the meat ration 'lasts for only three evening meals', and 'cannot be made to go further'. On Tuesdays and Wednesdays Vere would instead cook 'a handful of rice, dodged up in some way with curry or cheese', though the cheese ration meant she had to use it sparingly. To 'make-do for Thursday, Friday, and part Saturday', she acquired soya-bean flour sausages from a dairy. Milk rationing 'has driven me nearly demented all the week', she further lamented. Though Vere expressed her gratitude to the government for the food they did get, which meant they never went hungry, her wistful longings for certain tastes are scattered throughout her wartime diary. It was 'very difficult to get any eggs... almost impossible' at several points in London during the war, and more than once

Figure 24. The ration book of one person, Mr Norman Franklin, and his weekly rations of sugar, margarine, 'national butter', lard, eggs, bacon, cheese and tea, as issued in 1942 during the Second World War.

a sign declaring 'NO FISH' would greet her at the local shops. After finding fruit in scarce supply, it was an occasion of 'great joy' when she 'managed to get a few eating apples', and it was a noteworthy event when visiting her family home in Birmingham that Vere was sold five 'over-ripe' and 'half-bad' oranges (any fresh ones being reserved 'really for children', given how rare they were) from which she could at least extract the juice.[21]

Unsurprisingly, the government urged Britons to make the most of their food rations. To take just one example, one trick they offered – still used by keen bakers today – was to save the packaging once all the butter or margarine had been scraped off the paper and use it for greasing bowls or tins.[22]

Doctor Carrot and Food Substitutes

The government also recognised that rationing and food shortages meant coming up with crafty ways of substituting one food for another. A sweet tooth, the Ministry of Food decided, somewhat optimistically, could be satisfied with the natural sugars found in vegetables, especially carrots, which could easily be grown at home. In a video documenting celebrations for Easter 1941, two little children, socks up to their knees, holding hands trepidatiously, visit a store only to find the usual ice creams have been replaced with carrots on sticks, selling for one pence apiece. 'They're awfully good for you,' encourages the narrator lightheartedly, though it's not clear if the children's faces as they lick the solid vegetable are scrunched up in concentration as they savour the taste, or in disappointment.[23] 'As many a wise mother knows,' according to the Ministry of Food's cookery leaflet of 1941, 'the child who eats raw carrot freely is most unlikely to have a craving for sweets.'[24] The Ministry created the character 'Doctor Carrot', a bespectacled carrot with top hat and briefcase labelled with 'VIT-A' (a reference to the high vitamin A content in carrots). Doctor Carrot was to skip across many a Second World War leaflet and poster in an attempt to encourage carrot-eating, especially among children (see Plate 16). In a second, later, version of the Ministry's carrot cookery leaflet, carrot was used to replace sugar in a carrot cake, and in another alternative dessert called 'Carrot Charlotte':

8 oz. carrots, diced or sliced

Orange or lemon essence to flavour

1 tablespoon syrup

A few sultanas

1 oz. suet or margarine

4 ozs. breadcrumbs

Cinnamon to taste

Cook the carrots with the orange or lemon essence, syrup and sultanas, in a very little water. Cook until water is absorbed. Mix suet and breadcrumbs together. (If using margarine, melt and mix in crumbs.) Add cinnamon and press part of the mixture into a greased pie dish. Add some of the carrots and repeat layers to fill dish. Bake in a moderate oven for about 40 minutes. Turn out and serve with custard sauce.[25]

Much like carrots, potatoes were a saving grace during wartime as they could be produced relatively easily in gardens on the Home Front, then boiled, mashed, or fried in a variety of dishes. Moreover, potatoes are a good source of carbohydrates, and are highly nutrient dense, especially rich in vitamin C. As the food historian Rebecca Earle notes, 'potatoes are an exceptionally efficient way of converting sunlight, soil and water into nourishment', a hectare of potatoes producing enough protein to feed seventeen people for a year, whereas a hectare of wheat would only sustain seven.[26] When the Ministry of Food was first set up in 1916, it quickly learned the value of the potato in the war effort. Janet Lane-Claypon, later an acclaimed epidemiologist, was employed by the Ministry to test potato recipes, and ruled favourably on potato bread, in which potato was used in place of a portion of wheat, by then in short supply. Receiving sample loaves from the Ministry, the newspapers were in turn largely impressed with the culinary innovation. A letter sent to the *Liverpool Echo* on 12 January 1918 by a certain Hugh R. Rathbone included a recipe for potato bread:

> 9 lb. of flour, 3 lb of boiled and mashed potatoes, 1½ oz. barm*, four to five pints of water, salt to taste. Mix the potatoes in a basin with the barm and two or three handfuls of flour with a little water into a sponge. Leave it to rise for 15 minutes. Knead it with the rest of the flour and a little salt. This mixture properly baked should produce not less than 15 lb of bread.

If this were to be taken up by the nation, Hugh wrote enthusiastically, it would save eight million sacks of flour a year.[27] 'I make your rations go further,' declared another of the Second World War Ministry of Food's beloved anthropomorphic food characters, Potato Pete (see Plate 15), alongside a recipe for 'Fadge', or 'Irish Potato Bread'. With a couple of teaspoons of baking powder, and a dash of the milk ration, the flour/potato mix could make scones, with the option to add a little sugar for a sweeter, treat version.[28]

Housewives seeking clever ways to create mock forms of their favourite dishes, in the absence of all the usual ingredients, were inundated with tips and tricks from leaflets, recipe books and radio shows. Mock fish cakes could be made with a mix of potatoes, beans and anchovy paste, for instance, or mock sausages from lentils. Instead of the much-coveted turkey for the

---

* Foam skimmed from a fermenting liquid, used to leaven bread.

Christmas of 1941, listeners to BBC radio's *The Kitchen Front* were told to try 'murkey' instead. Comedy duo Gert and Daisy (played by sisters Elsie Waters and Doris Ethel Waters) described how to fashion the mock turkey by stuffing a leg of mutton with sausage meat, serving it with bread sauce.[29]

Perhaps one of the most notorious culinary losses for Britons during the Second World War was the banana. Tapping into a widely felt sentiment, in 1943 Harry Roy & His Band came out with the latest in a popular sequence of light-hearted swing hits (including the rather suggestive 'My Girl's Pussy'), entitled 'When Can I Have a Banana Again?'. The lyrics lament the wartime scarcity of sugar, tea, eggs, bacon and chicken, but communicate a longing for the taste of banana above all else. The type of banana that was craved across the country was not the standardised seedless Cavendish variety that entirely dominates worldwide banana sales today, but the larger, sweeter and creamier *Gros Michel* (literally 'Big Mike') banana. The *Gros Michel* was widely sold before the 1950s until the crop was ravaged by a fungus called Panama disease, rendering it virtually extinct.* It comes as no surprise, then, to learn of wartime attempts to create 'mock banana', using mashed parsnips and a touch of bottled banana extract. Yet, after longing for and dreaming about bananas, Joan Stokoe remembers her mother's mock banana did not live up to her expectations: 'It was awful!', she declared.[30]

By the end of the war, dried banana powder had appeared in a further attempt to satiate the nation's yearning taste buds. The British public were used by then to consuming food that had been preserved by removing the water content. Eggs had been rationed to only one per person per week from 1942, and their shortage during the Second World War led to that notorious invention:

---

* The Cavendish banana is also now under threat from the Panama disease as a symptom of our modern reliance on monocultures. Resistance to the wilt cannot occur if there is so little diversity in the crop.

tins of dried powdered eggs. Developed by the British Ministry of Food's Department of Scientific and Industrial Research, spray-drying eggs into a powder saved 20 per cent of the shipping space required for fresh eggs, while increasing their shelf-life (some estimates put it at between five to ten years). Now, everyone in Britain was entitled to one tin or packet of dried eggs (supposedly equivalent to twelve fresh eggs) every two months, in addition to their solitary weekly real egg.[31] The Ministry of Food published a handy booklet advising its readers on how to adopt dried eggs into their diets. They could be reconstituted simply by adding two tablespoons of water to every one tablespoon of powder before beating out any lumps. The mixture could then be used 'exactly as fresh eggs', in dishes like 'bacon and egg pie', 'cheese pudding', Madeira cake, omelette, scrambled egg and the more 'economical' version of the latter named (for some reason) 'English Monkey', which was made by soaking one cup of stale breadcrumbs in a cup of milk. One tablespoon of margarine provided the cooking medium for this breadcrumb mixture, to which half a cup of grated cheese and the equivalent of one reconstituted egg was added. Once scrambled the thrifty dish could be 'spread on toast'.[32] Despite the Ministry of Food's enthusiasm, those who remember powdered egg were less gushing about it. 'The two words which still make my blood run cold, are DRIED EGG,' declared Jill Beattie, who was a child at boarding school during the war. 'A two inch block of hard scrambled egg oozing with water which saturated the half slice of so-called toast beneath it', made a repellent breakfast, Jill recorded. 'The TASTE – ugh!', she concluded in disgust.[33]

Powdered eggs were accompanied by tins of powdered milk, both of which were imported in large numbers from the United States as part of Roosevelt's lend-lease programme. So as to remain officially neutral in the war, from March 1941 America leased, rather than sold, supplies to any nation deemed 'vital to the defence of the United States'. So-called 'household milk', was made of skimmed milk, and specifically packaged for Britain's

Ministry of Food. Before the Second World War, skimmed milk had a foul reputation in both Britain and America as a waste product from the dairy industry, a drink – like whey and buttermilk – fit only for pigs or those who found themselves in extreme poverty. In Edwardian England, the skimmed milk left over after the cream had been removed was poured down drains or – writers feared – used to adulterate 'real' whole milk.[34] Similarly, by the 1920s, many American country streams and fields were flooded with the festering dairy waste of nearby creameries which had little use for whey and skimmed milk.[35] New machines allowed milk to be dried at the turn of the twentieth century, and the appetite for creating new products from the protein (*casein*) found in wasted skimmed milk continued to grow over the following decades. During the First World War, for instance, casein extracted from leftover skimmed milk was

Figure 25. Tins of powdered milk and powdered eggs supplied by the USA for British civilians during the Second World War to supplement fresh rations.

used to coat aeroplane wings, and by the 1940s it could be used in paints, piano keys, plastics and even an artificial fabric dubbed 'milk wool' (though not without leaving behind a sulphuric acid-contaminated by-product).[36] Yet it was only with the onset of the Second World War, and the need to supply hungry allies to support the war effort, that skimmed milk moved from hogwash into mainstream British diets as a nutritional source of human food. Powdered milk cost 9 pence when Vere Hodgson first found it in the local shops in late 1941. Along with flour and powdered eggs, Vere used powdered skimmed milk to bake a cake, which she admitted was really 'quite good'. Still, she wrote, '[I] shall be glad when the cows are doing full time again – I am more than ever in favour of cows.'[37]

On a larger scale, war opened up a global market for preserved, long-lasting foods. Created by evaporating some of the water content and replacing it with sugar as a preservative, tins of condensed milk first found use as a field ration during the American Civil War of the 1860s when they were supplied by Gail Borden's New York Condensed Milk Company. The First World War then reignited the market for tins of condensed milk, by which time Nestlé dominated in their production, fulfilling government contracts across Europe. Another tinned product, canned tuna, emerged on the British market for the first time in the First World War. Imported from the United States, it was branded as 'Jack Tar Tuna Fish', a new cheap protein source that tasted 'like the breast of tender chicken', and that because it was soaked in oil would help the consumer 'over the butter shortage too': 'an absolute treasure in War-time'.[38] In the Second World War, the British were introduced to American-made Spam, a now-iconic budget brand of canned cooked pork shoulder, made from scraps of this less desirable cut of meat that had been ground into a paste, mashed up with water, potato starch, salt, sugar and the preservative sodium nitrate, and formed into a square-shaped pink lump. Spam made its debut in 1937 from Minnesota, and was

A Delightful
Surprise.

Everything is *so* dear and
you don't know *what* to get
for a change.

Here is something entirely
new—quite different from anything that has ever been sold in
this country before.

It is "Jack Tar Tuna";
a delicious fillet of fish; tender and delicate and tasting
like chicken.

Ask your Grocer for a tin.
You will be delighted with
this new fish food.

Imported and guar-
anteed by the pro-
prietors of Sailor
Slice and Skippers.
If unable to get it,
send your Grocer's
name and address
and 1/- in stamps for
a quarter-lb. tin to
ANGUS WATSON &
Co., Dept. D. 16,
Newcastle-on-Tyne.

**JACK TAR
TUNA FISH**

Figure 26. An advert for canned Jack Tar Tuna Fish, presented as a
new way of getting around the food shortages of the First World War.
The Daily News, Friday 25 January 1918.

marketed towards housewives in search of a quick dinner option,
but it found its fortune as a wartime food, distributed to the US
military and for consumption on the Home Front as part of the
lend-lease agreement. Margaret Thatcher, prime minister from
1979–90, reportedly referred to Spam as a 'wartime delicacy'.
Soon after the war, meat shortages led the British government
to seek out alternative products, like canned snoek, a smelly
fish brought all the way from South Africa. This proved too
adventurous for British palates, however, and much ended up
wasted as cat food.

Beyond these tinned foods, there were further advances in food-saving technologies brought about by global war. The US developed a new technique to ship meat during the Second World War, called de-boning, in which the bones, fat and worst-quality cuts were stripped from carcasses, which were then compacted down into 50-pound containers. Adopted by New Zealand's meat industry in their wartime exports to Britain, de-boning meant that the meat took up 60 per cent less space in the refrigerated ships than if it had been stored as carcasses.[39]

Salvaging Scraps

On the Home Front, the thrifty cook would be sure to use up any leftovers to make precious food stretch across multiple meals. Stale crusts, for instance, could be crushed and saved to coat foods for frying, used to make stuffing, or even an indulgent 'crumb fudge', according to one Ministry of Food booklet published in the Second World War. The recipe for the latter is as follows:

    2 tablespoons syrup

    2 oz. margarine

    2 oz. sugar

    2 oz. cocoa

    Few drops vanilla, peppermint or orange essence

    4–6 oz. dried crumbs

Heat the syrup, margarine, sugar and cocoa gently until all is melted. Stir in the required flavouring and then the breadcrumbs. Mix thoroughly and turn into a well-greased 7 in. sandwich tin; spread evenly and mark lightly into fingers or squares. Leave for 24 hours and then use as a cake or sweet. This fudge improves with keeping for a day or two.[40]

The Ministry of Food also recommended cooking potatoes in their skins so as not to waste them, using up the potato water in soups and gravies, and recycling any leftover potatoes into pastry, pancakes or potato salad.[41] What's more, the remains of mashed potato could be combined with offcuts of cooked fish to make croquettes or 'fish pudding' (presumably similar to what we might call 'fish cakes'). With a bit of grated cheese or minced onion, old mashed potato became a quick and tasty sandwich filling 'for the midday lunch-box', according to Potato Pete.[42] Meanwhile, to help stretch the butter ration, leftover fat from meat was to be rendered down by adding water and letting it evaporate until the fat could be strained. The small brown chunks that often remained in the strainer after this process were also to be reused, this time in pies or stews.[43]

In the Edwardian era waste had come to be seen only as something to be disposed of rather than reused or recycled, but the shortages of war dramatically shifted the value attributed to all types of kitchen rubbish, including food scraps. In 1918 the National Salvage Council was formed, encouraging local councils to seek out rags, paper, bottles, jars, as well as tins and waste food. In the Second World War, a salvage department was created within the Ministry of Supply, which encouraged – and by 1941 even compelled – local councils to collect salvage, making distinct collections for different types of rubbish including kitchen waste. Moreover, salvage centres, much like the old Victorian rag-and-bottle shops, were set up with the spoils of waste collection. Wearing an elaborate hat and fur-lined coat, the Queen consort (Elizabeth, wife to George VI, and mother of the future Queen Elizabeth II) visited a salvage centre in South-East London, along with the Mayor of Paddington, in 1941. The public broadcast of the event, which highlighted the value of the tins and bottles that had been collected and stored in particular, was part of a huge government-run publicity and propaganda drive aimed at reintroducing salvage into the public's consciousness and behaviour.[44] Those on the Home Front who saved and donated these otherwise useless items were praised for doing their bit for the war effort by reigniting the salvage economy. Even a discarded cereal box was worth salvaging; just one empty packet could be made into two practice targets for pilots.[45] As the head of the National Salvage Council said in 1942, 'the need for salvage is greater today than at any time. The housewife is just as important as the girl who makes the shell is just as important as the man who fires it, but unfortunately many have not yet realised it.'[46]

Kitchen leftovers that couldn't be eaten were also sought after for a number of war-related materials. The Council appealed for fruit stones and nut shells, for example, to be subsequently burned down to make the charcoal filters in gas masks. Old bones and fat were the subject of one poster from 1917, which loudly declared 'SAVE THEM FOR MUNITIONS' (see Figure 27). Glycerine, from the extracted fat, could then be used to make fertilisers, but also as glue for aeroplane manufacture, and even explosives. Rich in phosphorus needed for soil health, discarded bones were actually a vital 'weapon of war', as one poster, this time from the Second World War, put it to the public. Residents were again encouraged to hand over bones to be gathered by a special bone collector or the dustman.[47] Such campaigns were not unique to Britain. In the US, Walt Disney's Minnie Mouse and Pluto were conscripted for a film that encouraged housewives to save leftover fats from cooking to be used to make explosives.

This choice of applications highlighted a continual wartime conflict between supporting agricultural yields and the munitions industry. In a process first used on an industrial scale in 1913, the German chemist Fritz Haber developed a new way of 'fixing' nitrogen from the atmosphere into liquid ammonia. His invention was hugely significant in the context of the First World War, when the British blockade left Germany unable to import natural nitrates from Chile, then the world's biggest exporter.[48] The Haber-Bosch method (named after Haber and his assistant Carl Bosch) of fixing nitrogen made way for the creation of synthetic fertiliser. It is still the main method of nitrogen fixing in today's synthetic fertilisers that feed nearly half of all the crops we eat, and according to the science writer Charles C. Mann, for this reason 'more than 3 billion men, women and children – an incomprehensibly vast cloud of dreams, fears and explorations – owe their existence to two early-twentieth century German chemists'.[49] Fritz Haber was awarded the Nobel Prize for his work in 1918. Yet the process of nitrogen fixing is also key in the creation of explosives, and nitrogen was routinely rerouted for use in munitions manufacture in the Second World War.

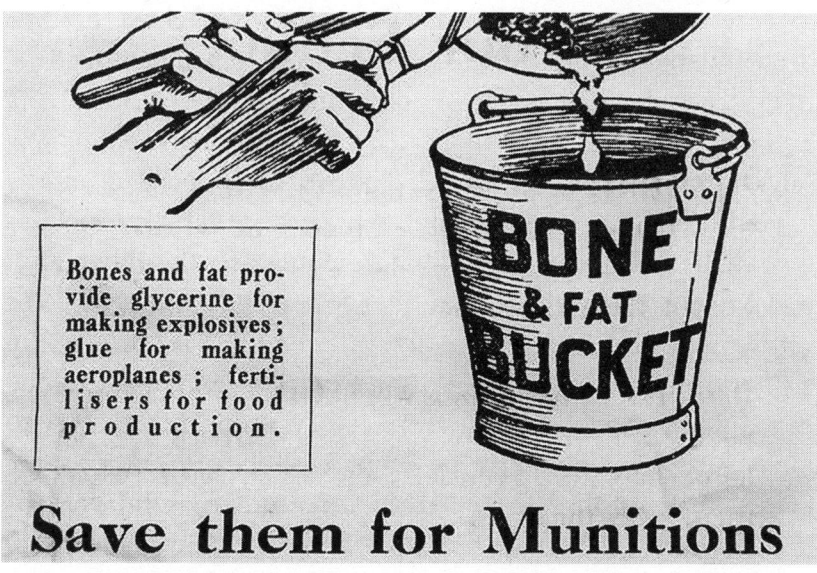

Bones and fat provide glycerine for making explosives; glue for making aeroplanes; fertilisers for food production.

BONE & FAT BUCKET

Save them for Munitions

Figure 27. 'BONE & FAT BUCKET. SAVE THEM FOR MUNITIONS'. A poster printed by S. H. Benson Ltd, Holborn, with approval of the National Salvage Council and the Ministry of Munitions, 1917.

In another British campaign from the Second World War (see Plate 17), a winking pig waves with one trotter and rests the other on a bin labelled 'PIG FOOD'. 'We want your KITCHEN WASTE', the friendly swine seems to proclaim, encouraging the housewife to keep it dry and uncontaminated so that the council can collect it to feed pigs and poultry. The smelly task of emptying these food waste bins, one of which was normally shared by a community, was often taken over from municipal dustmen or at least dramatically helped by voluntary wartime organisations, especially the Women's Voluntary Service (WVS). Thousands of these uniformed women also took on the huge job of educating communities about the importance of salvage – kitchen waste as well as paper, rags, metal and rubber – distributing leaflets and going door-to-door to inform inhabitants of their new wartime value. Schoolchildren, too, were recruited as junior 'Salvage Stewards', or 'Cogs', to encourage the local community to save more waste for reuse in the war effort. The WVS helped the children understand their duties by handing out informative leaflets. In 1942, an official guide for school salvage stewards included this explanation of the children's duties: 'to multiply the amount of salvage collected in your school, to see to it that the boys or girls in your form, or in your House, know the value of their efforts, as well as the value of waste materials – paper, metal, rubber, rags, bones and kitchen waste'. This vital role, it said, would help the country 'nearer Victory'.[50]

Once collected, kitchen scraps were distributed to special farms or production plants, where they would be laid out to make sure that there were no lingering bones or other solids, before being cooked to remove pathogens. The solid product that emerged, and which was used for pig feed, was known to Londoners as Tottenham Pudding or Wembley Pudding on account of the locations of the waste-food concentrators, which emitted a pungent smell for miles around as they worked. This system was indicative of the successful war-time drive to reclaim value in things that would ordinarily have been thrown away. Even the remaining fats that weren't incorporated into the pig feed would not be wasted, but sent off to the soap factory.[51] Councils had to be careful, however, that any swill was treated with a sufficiently high temperature after outbreaks of foot-and-mouth disease damaged the pig population and the resultant meat supply.[52]

Figure 28. WVS salvage workers Winifred Jordan (left) and Kathleen Kent (the head of the local pig food collecting unit) emptying the contents of a kitchen waste bin into their trailer with its characteristic logo. The photograph was taken on a street in East Barnet, Hertfordshire, 1943.

Pigs are a problematic food source in times of food scarcity as they eat human food, rather than living off grass like sheep, cattle and other ruminant farm stock. In the nineteenth century, commercial pig keepers in the US and Europe had begun to raise pigs on cereals in place of food waste, as the price of grain dropped. The amount of specially grown animal feed imported dramatically decreased during the war to almost nothing, however, so pigs were fed on readily available kitchen waste once again, and in turn they became a hugely valuable war-time supply of protein. With this in mind, in 1940 the Small Pig Keepers' Council (SPKC) was established, and those who could were encouraged to keep their own pigs, filling the troughs with their kitchen scraps and sharing the duties of managing and cleaning the sties. In Pig Clubs, 4,000 of which had been estab-lished by 1943,[53] groups of between four and twenty-five people came together to raise a pig on their shared food waste. When it was fattened up, half the animal's meat was sold on to the gov-ernment to become ration stocks. The members of the pig club divided the rest of the meat among themselves, adding some extra protein and a tasty treat to their otherwise restricted and monotonous wartime diets. Preserved, the slaughtered pig pro-vided plentiful supplies of salted bacon, black pudding, lard and pork pies. One recipe from the Second World War encouraged the use of 'every bit of the pig', including 'scraps and oddments' such as the liver and sweetbreads. These could be minced and beaten together with bits of fat and lean pork, seasoned and bound with a dash of stock, then rolled up and covered with the inner fat, before being baked in the oven.[54]

As I have suggested, the practice of rearing pigs on leftovers has a long history, in both rural and urban environments. A cottager, recalling his childhood in the 1860s–70s, spent in the Hertfordshire village of Harpenden, remembered how neigh-bours would keep a tub near the back door. Collecting their household scraps, they would give this food to the pig keeper, who would then offer 'a portion of the liver or some part of the offal' in return when it came time to slaughter in November.[55] More organised Pig Clubs had also emerged in the late nineteenth century, for the distinct purpose of protecting the pig keeper from the premature death of their pig, each member paying a fee that insured the owner with a pay-out if anything were to happen to their animal. During the Second World War, government ministers took inspiration from these well-proven methods of saving food. The Pig Club scheme was supported by King George VI, who posed for a propaganda photograph with a rather chunky swine that had been fed on waste-food swill.[56] In a morale-boosting poem from 1941, the Ministry of Food suggested that housewives who saved kitchen waste to reuse as pig feed were helping Britain win the war:

Because of the pail, the scraps were saved,
Because of the scraps, the pigs were saved,
Because of the pigs, the rations were saved,
Because of the rations, the ships were saved,
Because of the ships, the island was saved,
Because of the island, the Empire was saved,
And all because of the housewife's pail.

Indeed, by 1943 31,000 tons of food waste were being saved every week, enough to sustain 210,000 pigs.[57]

Any kitchen waste that could not be fed to livestock might instead be made into nutrient-rich compost. According to the Ministry of Agriculture, which even provided a handy diagram to explain how to build a compost heap, 'pea or bean or potato haulms', meaning the stalk or stem, as well as the 'outer leaves or tops of vegetables' should be combined with leaves and other cuttings from the garden to make the first layer. This would be alternated with a layer of soil, as well as a bit of animal waste acting as the 'starter' to ensure decomposition took place. Kitchen waste would then become a 'valuable manure', which would be fed back into the earth to help ultimately in the national drive for food.[58] Whether they were reused in thrifty recipes, pig feed, fertilisers and weapons of war, or finally simply composted, leftovers truly did help to win the world wars.

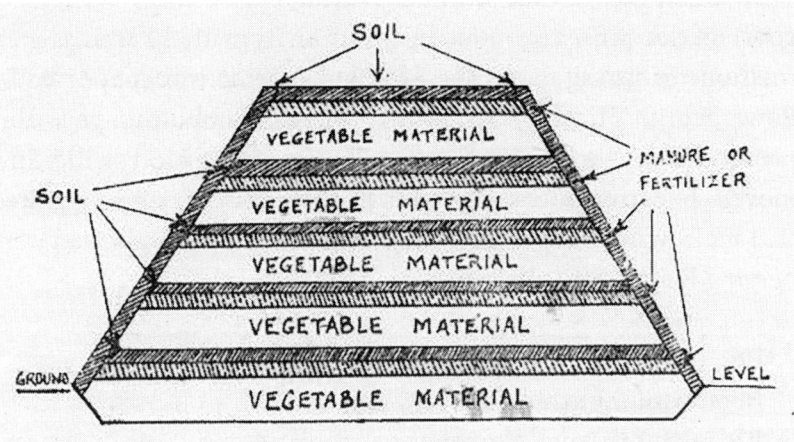

Figure 29. 'How to Make a Compost Heap', according to the Ministry of Food's Dig for Victory Leaflet Number 7, 1943.

Dig for Victory

> The battle on the kitchen front cannot be won without help from the kitchen garden. Isn't an hour in the garden better than an hour in the queue?
>
> Lord Woolton, Minister of Food, 1941.

Life on the Home Front meant avoiding as much food waste as possible and putting any food that could not be eaten to good use elsewhere, but this did not alone account for the dearth of supplies;

it was also necessary to increase the amount of food produced within the country. From 1917, the 'plough-up' campaign was managed by the County War Agricultural Executive Committees (CWAECs), which sought out land to expand the domestic production of staple crops like wheat and potatoes. Land under cultivation increased by 19 per cent over the course of the First World War to 5.01 million ha.[59] Allotments, railway scrubland, sports fields and golf courses were all co-opted for this new mission. Famously, even the king (George V) ordered that the flowerbeds at Buckingham Palace be transformed into vegetable patches. Then, in the late 1930s, as a second world war became inevitable, farmers were compensated for ploughing up grassland and planting it with wheat, oats, barley, potatoes and other core crops. This work was essential given that food imports dropped dramatically during the war from around 22 million tons per year to between 11 and 15 million tons.[60]

Meanwhile, in the 'Dig for Victory' campaign of the Second World War, ordinary people were encouraged to grow their own fruit and vegetables. Detailed leaflets were issued offering the novice gardener advice on a huge range of agricultural topics: how to dig effectively, how to plant different types of seeds, how to store root vegetables, how to preserve fruits and vegetables into jams, and how to dry, salt or pickle them. How, too, might the home vegetable grower fight off the enemy when it came in the form of peckish creepy crawlies? As green-fingered readers will know, cabbages alone were – and are – habitually munched on by slugs, flea beetles, cabbage root flies, caterpillars (of cabbage white butterflies and cabbage moths), cabbage aphids and the cabbage white fly. As the Dig for Victory leaflet declared, to lose crops to these ruthless pests 'in war time, when every ounce of home-grown food is needed' was a 'serious waste'. The gardener was therefore, it advised, to become familiar with pesticides, including nicotine, naphthalene, metaldehyde and pyrethrum. Many of these were the result of nineteenth-century chemical research; only the last is still allowed to be used today in insecticides based on pyrethrins, the others being banned on account of their toxicity.[61] As part of the Dig for Victory initiative, Village Produce Associations (VPAs) were set up in rural areas, encouraging members to bulk buy seeds and fertilisers, and to trade any surplus agricultural outputs among themselves. By 1943 Britons up and down the country were, collectively, successfully growing over a million tons of vegetables in their gardens and allotments, thanks in great measure to these informative drives to 'Dig for Victory'.[62]

Agricultural labour had declined in the years before the First World War, as women and men moved from rural areas to the towns and cities, and, as we saw in Chapter 4, increasingly food was imported from abroad rather than home-grown. As young men were called up to fight, by 1915 authorities started to look for female labourers to work on farms, where crops needed to be tended and harvested to prevent them and valuable agricultural land from going to waste. In March 1917 the Women's Land Army was officially created, its mandate neatly explained by one of the organisation's songs:

> The children shall not starve,
> The soldiers must have bread,
> We'll dig and sow and reap and mow,
> And England shall be fed.[63]

At first, the idea of female agricultural labour was met with some resistance. Annie Sarah Edwards, known affectionately as Nance, was working as a domestic cook for a couple in the Surrey village of South Nutfield when news arrived that the First World War had broken out. With the help of Mrs Beeton's cookery book (still hugely popular a couple of generations after its first publication), and another paper-covered manual that her mistress kept, Nance had succeeded in producing simple, well-liked dishes

1. The animals whose meat was wasted at John Key's aborted banquet haunt his dreams in this satirical drawing. M.G., *Fatal Effects of Gluttony / A Lord Mayor's Day Night Mare*; lithograph, 1830.

2. The remains of a feast, the half-eaten pie and broken wine glass a symbolic chastisement of wastefulness and intemperance. Willem Claesz Heda, *Banquet Piece with Mince Pie*; oil on canvas, Dutch, 1635.

3. A lavish baroque feast, showing swan and peacock pies and sugared sweetmeats. Peter Paul Rubens and Jan Brueghel the Elder, *The Sense of Taste*; oil on panel, Flemish, 1618.

4. Salted cow's tongue and pigs' trotters were celebrated early modern dishes. Jacob van Hulsdonck's *Still Life with Meat, Fish, Vegetables, and Fruit*; oil on panel, Flemish, *c.* 1615–20.

5. Sugarloaf, a preservative in the early modern kitchen, shown partly wrapped in sugar paper dyed with indigo to repel insects. Unknown artist; oil on canvas, German, *c.* 1680.

6. A dog and cat fight over kitchen scraps. Pieter Cornelisz van Rijck, *Kitchen Scene with the Parable of the Rich Man and Poor Lazarus*; oil on canvas, Dutch, 1610–20.

7. A banqueting trencher decorated with the parable of Dives and Lazarus, a reminder to the viewer to donate their leftovers to the poor; wood, English, late 16th century.

8. By the Victorian era, artists romanticised gleaning as a feature of the rural past. Samuel Palmer, *The Gleaning Field*; tempera on mahogany, British, *c.* 1833.

9. A family makes use of a slaughtered pig, stuffing sausages, collecting its blood in a bowl, while children playfully blow up its bladder. Barent Fabritius, *The Slaughtered Swine*; oil on canvas, Dutch, 1665.

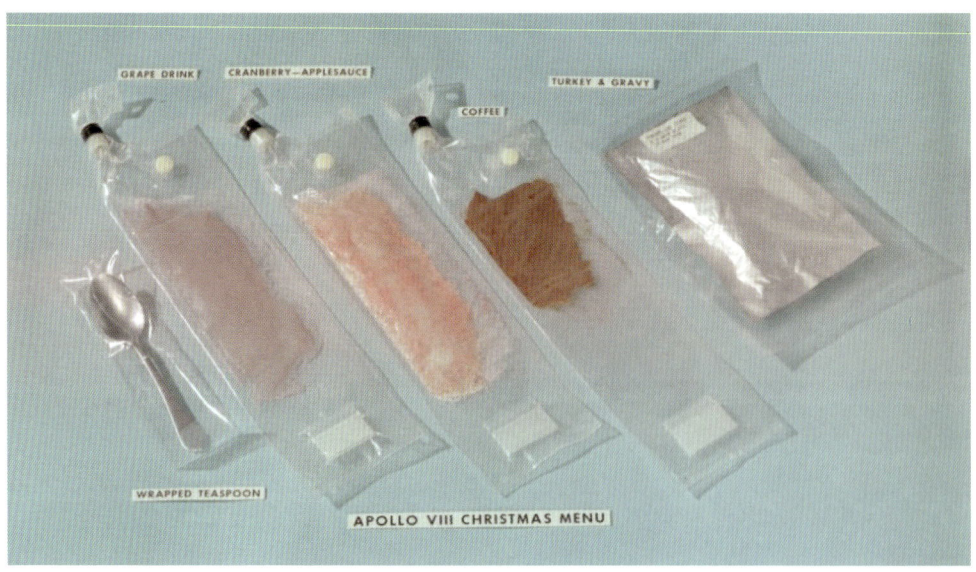

10. *The Pig*, from a series of lithographs entitled 'Graphic Illustrations of Animals. Showing their Utility to Man in their Employments During Life and Uses After Death', created by Benjamin Waterhouse Hawkins and published by Roake and Varty, London, *c.* 1850.

11. The Christmas menu aboard Apollo 8, including a 'wetpack' of turkey and gravy, a new technology invented for use in space. Photograph dated 7 March 1969.

12. 'Waste Not, Want Not' on a Victorian bread plate, designed by Augustus Welby Northmore Pugin (1812–52). The phrase was first found in print in the early nineteenth century.

13. From around 1900, an advert for Bovril, a mass-produced meat extract which was advertised as 'more nourishing' than homemade beef tea.

14. Brands advertised themselves as 'pure', 'unadulterated' or 'free from preservatives' in the light of the late Victorian adulterated food scandal, as in this trade card from London's Express Dairy Milk company.

15. *Potato Pete's Recipe Book* (London, 1943) was filled with thrifty recipes to reduce kitchen waste. Potato Pete was the invention of the Ministry of Food in the Second World War.

16. 'DOCTOR CARROT the Children's best friend'. A Second World War poster printed by Gilbert Whitehead & Co. in New Eltham, London, seeking to encourage carrot-eating, often in place of sugar.

17. 'We want your KITCHEN WASTE'. A Second World War poster by John M. Gilroy, encouraging the salvage of kitchen waste for pig food.

18. 'DIG ON FOR VICTORY'. A Second World War poster by Peter Fraser, as part of the wartime efforts to increase home-grown food.

19. FareShare distributing food that would have otherwise gone to waste at a Feeding the 5000 event in Bristol, 1 June 2013.

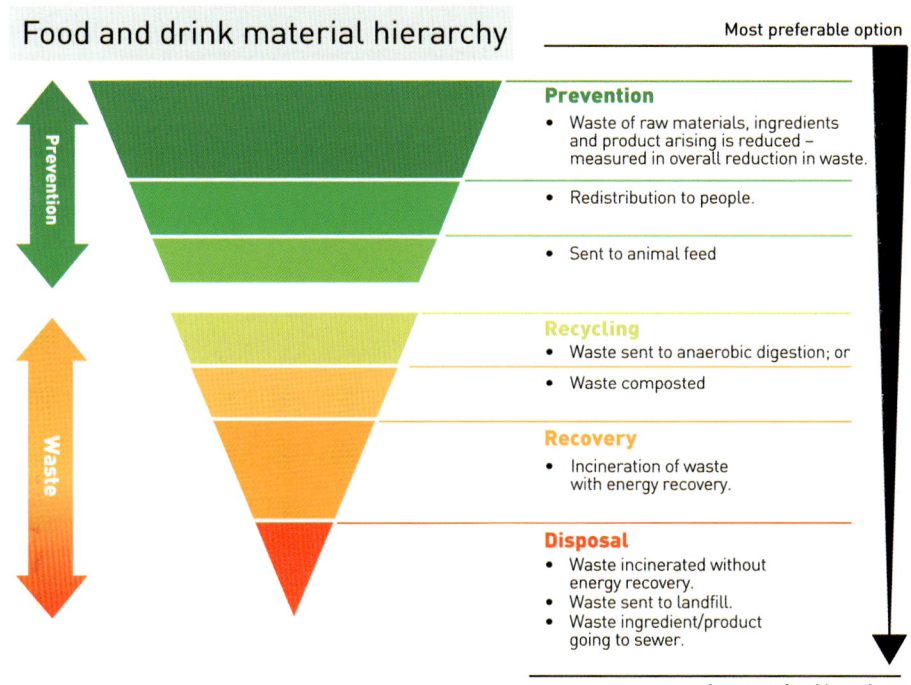

Food and drink material hierarchy

Most preferable option

Prevention
- Waste of raw materials, ingredients and product arising is reduced – measured in overall reduction in waste.
- Redistribution to people.
- Sent to animal feed

Recycling
- Waste sent to anaerobic digestion; or
- Waste composted

Recovery
- Incineration of waste with energy recovery.

Disposal
- Waste incinerated without energy recovery.
- Waste sent to landfill.
- Waste ingredient/product going to sewer.

Prevention

Waste

Least preferable option

20. WRAP's Food and Drink Material Hierarchy, showing 'prevention' of waste as the preferred option with 'recycling' and 'recovery' before 'disposal' as the last option.

21. Callum Pearman, a student at Durham University, was one of the seasonal workers who harvested cherries in Canterbury in summer 2020 to stop them from going to waste.

22. Modern technologies help to prevent the natural imperfections and insect burrows shown in Giovanni Battista Recco's *Still Life with Apples, Cabbage, Parsnip and Lettuce*; oil on canvas, Italian, *c*. 1615–60.

for her employers. She had grown up on a farm in Nutbourne, Chichester, feeding the pigs acorns, crab apples and meal from the local windmill, brining their meat at slaughter time and scaring off rooks from pecking at the corn in the field. Attracted to the prospect of returning to the land and putting her agricultural skills to use, Nance signed up to the Women's Land Army, and sought out three character references as was required by the organisation in its early years. Hesitant to give her blessing to the request, the local canon's wife was uneasy about the shift in gender norms that working as a 'Land Girl', as they were known, would entail. '"Do you know what it involves? It means that you'll be dressed as a man! And I object to that,"' she said, continuing '"it's a disgrace to show your ankles".' Nance agreed: 'Of course it was, and I had long skirts to [...] the ground in those days.' Luckily for Nance, who was eager to play her part in the war effort, the canon intervened to assuage his wife's fears, maintaining that since Nance was a good girl the breeches were unlikely to 'alter [her] life'. Looking back, Nance was quite sure that her experience didn't affect her morals, but it did alter her clothing. After long days working with horses, ploughing and growing crops, Nance ended up dispensing with her steel corset, an essential in women's fashion of the day, which she remembers rusting from her sweat.[64] As the shortfall in agricultural labour became more apparent, the Land Girls slowly found more acceptance on farms.

Initially, recruits were taken from the middle and upper classes, but by 1918 the Women's Land Army was accepting a wider range of volunteers to help make sure crops were harvested and food supplies adequate. Many of these women had little experience in agricultural labour, and the Board of Agriculture therefore supplied training in farm work – milking, butter and cheese making, feeding animals, cleaning out enclosures, hoeing, harvesting – normally for four to six weeks. Doris Heap (later Robinson) remembered having spent just two weeks training before being sent to a farm in Loughton in Essex, where she was in charge of seven Jersey cows, hundreds of hens, a goat and some unruly ducks. Her days were long, starting at six in the morning with milking and sometimes not ending until 10 p.m., given that the hens' eggs in the incubators needed to be turned. She remembers the farmer being 'a bit of a strange man', who was very strict, listening out for her footsteps on the gravel in the morning and scolding her if she were as little as five minutes late. He never allowed her a day off. At Christmas 1917 Doris was surprised to hear that the farmer would be giving her a Christmas present, but her delight at this supposed gesture of seasonal goodwill quickly vanished when she learned it was 'half a dozen mouldy pears' that the farmer wanted to clear out of his larder! Still, she found joy in tending to the animals and knowing that she was helping in the war effort.[65]

Most Land Girls in the First World War were stationed by themselves on farms that requested help. Some, however, went from farm to farm in groups where labour was needed temporarily, depending on the season. Known as 'threshing gangs', they helped in the laborious physical task of manually threshing grain, loosening the edible part from the straw. Rosa Freedman was one of these mobile Land Girls, who worked on a market garden picking frosted sprouts in the winter and pulling up rhubarb in the spring. Her gang were also employed to pick fruit – raspberries, gooseberries, blackcurrants, plums and apples, 'climbing tall trees with baskets tied around our waists' – at Dumpton Park, in Broadstairs, Kent. In their war work, no food that could be salvaged was to go to waste. Indeed, Rosa remembers collecting over-ripe fruits to send to a jam factory.[66]

When the Second World War loomed into view, the Ministry of Agriculture, fearing impending food shortages, approached Lady Gertrude Denman to re-form the Women's Land Army. This she accomplished in June 1939, three months before Britain declared war on Germany. Advertisements soon appeared in the newspapers, encouraging women to sign up to keep the farms

Figure 30. Land Girls harvesting beetroot and tractor-driving in 1942/43.

running and the nation fed in the (increasingly likely) event of war. One of the early recruits who replied to a newspaper coupon was Margaret Mary Rumbold, who in the summer of 1939 had just turned eighteen, and was working long hours as a general dogsbody in a small boarding house in the beach-front town of Folkestone in Kent. When war was declared, Margaret was promptly called to Wye College, where the Women's Land Army trained recruits. A self-professed 'tom boy' and a 'grafter', who had spent many hours as a child 'knocking about' on her friend's farm, Margaret thought herself more prepared than many of the other women for the harsh physical labour and long hours that farm work entailed. She recalled fondly being called a liar by one of the tutors at Wye College, who couldn't believe that she hadn't milked a cow before: 'But I hadn't milked cows before. But I had watched, many times, the cowman at my friend's farm, milking cows. So yes, I did know what it was all about, and it's like many other things, if you watch for long enough you must get some idea, mustn't you?' Still, many of the countryside tasks came as a shock to those who had only lived in towns. Margaret recalled how the tutor 'tells you to get the udder washed. Well, we're town women, and we don't know too much about cows, and that horrible bag thing underneath you don't reckon much about it […] but you get down to it because he tells you to hurry up time's getting on.' Some women were so exhausted they almost crawled home each evening: 'You are hoeing fields of onions, little onions, all day long, day after day, with that hoe, and you've never done it before, you've been at a typewriter or something, you can imagine the physical state you're in at the end of the day.' All this under the dark spectre of war. Margaret remembers one day picking Brussels sprouts with a woman named Gladys, when German planes appeared overhead. Gladys suddenly disappeared:

> I said has anyone seen Gladys, you know, and you're looking round […] and there she is spread out on the floor on the ground, and because Brussels sprouts grow about 4 foot high in fields […] and she got a whacking great Brussel leaf over her head, and I said do you feel safe under there Gladys […] she said I know it's stupid but as long as I've got something on me head, she said, I don't care what happens. So she felt happy down there, while they were going over [laughs].[67]

Land Girls, like many other agricultural workers, received an additional cheese ration each week, which could easily be packed up to eat in the fields. Margaret also remembers an extra fat ration during the harvest season, the one time of the year when 'everybody went mad' for pastry. Like many Land Girls, however, Margaret recalled living with a miserly mistress, who never let her have more than her rations. The dairy farm had plenty of milk, and although it was collected by the Land Girls, they were entitled to none of it if they didn't live on the farm and were billeted elsewhere. Margaret's landlady made much of it into butter and sent it illegally to her friends who were struggling on the meagre allowance provided by their ration books alone.[68]

These are just a few of the stories of the roughly 23,000 women who volunteered to work in the Land Army during the First World War,[69] and of the over 200,000 women employed as Land Girls during the Second World War, when (from December 1941) the National Service Act legalised women's conscription into domestic war work. By mid-1943, 80 per cent of married women had taken up war-time jobs and 90 per cent of single women.[70] Even in the First World War, women worked in a variety of jobs in the food industry, determined to keep food on plates. In one photograph from 1918 (Figure 31), four women pose with bread in a bakery made in their role within the Queen Mary's Army Auxiliary Corps. The Women's Army Auxiliary Corps (WAAC), as it was first called, was formed in 1917 to free up men to fight in the war; they worked in France and in the UK.

Figure 31. Women from Queen Mary's Army Auxiliary Corps working in a bakery in Dieppe, France, 1918.

When the sharp stab of hunger was beginning to be felt in Britain in 1914, another women's organisation was set up, which would help to win the war through its food initiatives: the Women's Institute (WI). Madge Watt had been a member of the organisation in her native Canada. Immigrating to Britain, Madge saw the potential for a war-time version to be set up here, and secured governmental funding to do so on 25 September 1915. The initiative greatly expanded in the year 1919, when Lady Denman, the 'chairman' of the organisation and director of the Land Army, said of it: 'Through their initiative more food is being grown, help is being given on the land, food is being saved both by individuals and by united action, which results in the preservation of fruit and vegetables, the provision of dinners for school children and other co-operative efforts.'[71] In the Second World War, the WI continued to play a leading role in preventing food waste in the countryside, where they were stationed at least in every third village. The women who made up its members, who came from diverse social backgrounds, grew and sold surplus vegetables from their gardens at WI market stalls. To encourage others to grow more, they distributed seeds, especially of onions, which were sorely missed after the German blockade of 1940 saw continental imports come to an abrupt halt. They also educated others about the best way of making compost from food waste when fertilisers were in short supply. What's more, they helped to collect bones for munitions, set up Pig Clubs, and most famously they bottled, canned and otherwise preserved fruit and vegetables. Factories continued to make jam from farmed fruit, but the Women's Institute went to work collecting and preserving surplus fruit that would otherwise go to waste in the wild countryside, as well as from gardens and allotments.[72] These were skills passed down through generations of rural housewives who had for hundreds of years, as we have seen, ensured that the sweetness of out-of-season fruits could be enjoyed over cold winters. Now, the government called upon these women to volunteer their preservation skills for the sake of the nation's food supply. In 1940, the Institute obtained £1,400 worth of sugar from the Ministry of Food to distribute to their fruit preservation centres – often simply a village hall with limited facilities – where members made and packaged jam that was then sent to shops to be sold on the ration. In doing so, that year the Institute prevented more than 1,600 tons of foodstuffs from rotting.[73] The next year the Ministry struggled to supply extra sugar to the preservation centres, but urged countrywomen still to ensure that no fruit was wasted on trees and bushes. Once gathered it could be taken and sold to one of the 4,500 centres run by the Women's Institute, which, as Lord Woolton put it, were 'ready with their workers, ready with their jam jars, and the covers for them; everything ready for your help in bringing in the fruit'.[74]

———•———

'Every war has had immense and far-reaching influence on every nation's kitchen history, but never have any women at any time had a grander opportunity to influence the kitchen history than the opportunity offered the Land Girl to-day', declared cookbook writer Moira Meighn in the Land Army's monthly magazine in August 1940. These women were not only to help grow, tend, harvest and preserve the food that nourished the nation, but to change the food habits of the agricultural communities in which they stayed. In particular, Meighn argued, it was vital to persuade others not to discard the peel of root vegetables, where many of the most important nutrients were hidden, while 'throwing away the water in which greens have been cooked is tantamount to throwing away the baby with the bath water'.[75] As this chapter has made clear, Allied victory was not just ensured by the countless men who bravely risked – and lost – their lives in combat, but those people, especially women, who were left on the Home Front. At home, food was the most powerful weapon of war. Housewives, entreated to make less food go further by wasting less of it, repurposed leftovers and created palatable foods with available substitutes. Food was grown wherever possible, and young women were called up to carry out the back-breaking work needed to harvest and preserve Britain's farmland and its outputs. Meanwhile, campaigns sought out inedible waste products from eating, like bones, tin cans and other food packaging to be collected and reused for munitions, fertiliser and other war-time wants. The salvage efforts of the Directorate of Salvage and Recovery alone over the ten years between 1940 and 1950 saved more than 8 million tons of bones, kitchen waste, textiles, paper and scrap metal from households.[76] Through the collective action of ordinary people, and an effective rationing system, Britain survived this period of global war without having to face the devastating effects of mass starvation.

However, Britain's success in feeding its people (and indeed in fighting in combat) was reliant on its empire and the various dominions, (ex-)colonies and protectorates that were still politi-cally tied to it. In the Second World War, Australia and New Zealand, for example, provided meat which was transported across the oceans in refrigerated ships, or in dehydrated or canned form. Tins of powdered milk and eggs (and Spam!) were just the tip of the iceberg in terms of the amount of food shipped over from the United States as part of the lend-lease system, which in total provided a staggering sum of goods amounting to roughly $30,000,000,000.[77] Addressing the colonies, Britain's secretary of state requested that subjects do away with any unnec-essary consumption and reduce food imports. The food historian Lizzie Collingham has drawn attention to the trenchant response of the resident commissioner in Bechuanaland (a British-ruled protectorate later to become the Republic of Botswana), who pointed out that people across Africa were already surviving off a basic subsistence diet.[78] Most shockingly, the people of Bengal in British India were to suffer an appalling famine that ultimately caused over three million deaths. When local British rulers failed to effectively manage existing food stocks, they asked for help from the British government, but Churchill's cabinet refused to redirect food meant for Britain's inhabitants or its military, and even cut shipping to the Indian Ocean, believing Bengal's inhabitants less worthy of the limited provisions than those at home.[79] The wartime fate of Bengal exemplifies a recurring theme in the story of food waste and preservation: the unequal distribution of surplus and want.

Britain had spent the past century enjoying the economic benefits of its colonial empire, but after years of world war, the global power dynamic was set to change. When the dust started to settle, it was America that emerged as the strongest post-war economy. The lend-lease system had been designed to allow the US to support Britain and its allies without entering the war, but in December 1941, after the Japanese attack on Pearl Harbor, America felt compelled to join in the fight against Nazi Germany and the Axis allies. Unlike any other combatant nation, the US was able to expand economically during the Second World War, to the point that it accounted for nearly two-thirds of the world's industrial production. This economic success translated naturally into cultural power. US-made cans of food supplied to Allied armies were found across the globe, along with cans of the now-ubiquitous Coca-Cola, which soon became desirable to the Europeans who saw them being consumed by US military personnel at their army bases.[80] As Cold War tensions burned, America actively sought to impose its culture on the rest of the world. The Marshall Plan, which saw $13 billion (equivalent to around $114 billion in today's money) injected into Western Europe's recovery from 1948, was granted only on the condition that each country be exposed to propaganda that encouraged the American way of life. While European stomachs still rumbled, the Americans constructed a convincing and enticing narrative of America's alimentary abundance. For those in the West who adopted the mass-consumerist, capitalist way of life peddled by the United States, home luxuries like refrigerators and televisions were now within reach. After years of war and worry about making do with limited food supplies, Europeans wanted a seat at the feast that the US was seemingly hosting. Government subsidies that had helped local authorities in Britain to recycle food waste, and that were continued under the post-war Labour government, were abolished along with other wartime controls in 1954 as British attention turned instead to building a free and prosperous consumer society. At the close of the first half of the twentieth century, then, the United States would take on an economic, political and cultural hegemony that would shape Britain's culinary culture in the second part of the century and beyond.

6

McDONALD'S, PUNKS AND FOOD ANARCHISTS

Macaroni and Cheeseless[1]

Makes: 90 servings
Need: 40 qt. pot, large mixing bowl, 3 12" X 18" baking pans
Preheat oven: 350 degrees
Prep time: 1 hour, 30 minutes
Baking time: 30 minutes

Elbow Macaroni

8 gals water
5 TBSP sea salt
20 lbs elbow macaroni (soy semolina)

Bring the water and salt to a rapid boil in a 40 qt. pot and add macaroni. Return to a boil and cook for about 10 minutes. Macaroni ought to be al dente; do not over cook. Drain and rinse with cold water until all the macaroni is rinsed and cold and set aside.

Cheeseless

36 cups nutritional yeast
12 cups unbleached white flour
½ cup garlic powder
½ cup sea salt
4½ gals boiling water
6 lbs margarine
1 cup wet mustard

In a very large mixing bowl, combine nutritional yeast, flour, salt and garlic powder and mix well. Add boiling water, a quart at a time, using a whip to stir. Add margarine and mustard and mix well. Place macaroni in each of the baking pans. Cover with cheeseless sauce, being sure to coat each piece of macaroni. Sprinkle toasted sesame seeds or bread crumbs over top and bake in 350 degrees oven for 30 minutes or until it is hot and bubbling.
Serve hot. (This dish freezes well.)

C. T. Butler and Keith McHenry,
Food Not Bombs (1992).

As the wastefulness of the modern food system became increasingly apparent in the late twentieth century, US anarchist group Food Not Bombs led in redistributing leftovers from the food industry to make (vegetarian) meals for those who were going hungry.

———•———

'Ice lollies [...] gosh, we thought we were rich!' Ann tells me, remembering what it was like to have a fridge for the first time in her council-owned home in Chippenham, Wiltshire. There was a little compartment where you could store a slab of ice cream, which couldn't be kept too long, but which was all eaten up for Sunday tea anyway, she recalls. As they were still too expensive to buy outright in the early 1960s, Ann (born in 1945) and her sister Sue, four years her elder, rented their fridges from the local council for 2 shillings and 6 pence a week. This was not an insubstantial sum. Sue's husband Gordon was at the time making around £10 or £11 a week in his job at a machine shop.[2] Fridges had become commonplace in the United States in the 1930s, but in the UK fridge ownership was only at around 5 per cent in 1950. By the start of the next decade this had risen to 20 per cent, and by the early 1970s the majority of households had one.[3] Like most of the post-war generation in Britain, therefore, Ann and Sue grew up without a fridge. Perishables were instead stored in a larder with a cold stone slab on the floor. The larder might also have contained a 'meat safe' or 'food safe', a small wooden or metal cupboard with mesh sides to allow the air to circulate while keeping out insects. Bottles of milk, delivered daily to your door by the milkman, could be kept in a shaded, cold bucket of water in the warmer months, or lined up on the windowsill in cool weather, a stone on top of each to stop the birds from pecking through the tin-foil top.

Rationing still lingered on into the mid-1950s, sugar rationing ending in 1953 and restrictions on meat finally being lifted in 1954. 'Nothing was ever wasted,' Ann said of that time. If their mother had served up suet pudding for an evening dinner, for example, she would 'fry it up the next day for our breakfast'. 'That was lovely wasn't it', Ann remembers, and Sue agrees: 'That was our dad's favourite.' Pig-trotters, 'chaps' (pig cheeks or jaw) and even sheep's head (enjoyed by their mother, at least, if not the rest of the family) were all eaten so as not to waste any part of an animal, while potato peelings and any other waste that wasn't going to be eaten was fed to the chickens in the backyard.[4] Meanwhile, leftover milk made batter, or, when it all too quickly soured, was used up in scones and bread.[5] In between her part-time work at the local cinema, Ann and Sue's mother made three cooked meals each day – their father returning from the foundry for lunch – as well as attending to other household tasks like the washing and cleaning. Without refrigeration, shopping had to be done practically every day and leftovers used up quickly to prevent them going off.[6] It is no exaggeration to say that fridge ownership would change the lives of Britons – especially women – who were born in the post-war era.

By keeping food cold, fridges prevented it from going off and going to waste. 'On that exciting and delightful day when we first see our very own refrigerator in our very own kitchen', wrote Margaret Ryan in the Women's Institute magazine *Home & Country* in the 1960s, 'our minds almost certainly turn [...] to the joy of keeping milk, meat, stock and so on without the dread that the cautious sniff will meet the unmistakable smell of "going off"!'[7] Meanwhile a manual for Prestcold fridges (which would become Britain's largest fridge manufacturer) in 1959 boasted:

You will find your Prestcold is of infinite value. After a little time, you will wonder how you managed without it. Day in, day out, summer or winter, it keeps food fresh [...] wastage of food is eliminated. You can buy in large and more economical quantities without fear of the food deteriorating.[8]

Freezers took about a decade longer to integrate into the British household. The report of the National Food Survey between 1966 and 1970 recorded a 12.5 per cent increase in the purchase of convenience foods, most of which were quick-frozen goods, and by the start of the 1980s about half of homes had their own freezer.[9] A youthful Mary Berry, then named a 'leading authority on freezing', published *Popular Freezer Cookery* in 1972, which encouraged sceptics to make use of this newly available technology. Posing the question 'Why have a freezer?', Berry answered that freezing meant saving money, by buying food when it was at its cheapest or in bulk at a discount and preserving it for later use. Secondly, she wrote, 'You are released from the frenzy of work that ties you to the kitchen.' Housewives, and especially those who were working, could now prepare food in advance and simply reheat it (which meant, notably, that they wouldn't have to miss out on the conversation at a dinner party).[10] Indeed, a freezer owner no longer needed to use up leftovers from a joint of meat over the course of a week or to preserve fruits and vegetables in the traditional ways. Freezers could therefore slash a household's food waste output. As Mary Berry wrote in 1972, 'the remainder of a cake made for a special occasion can be put into the cold store', and 'a glut in the garden is food wanted not wasted'.[11] Indeed, Sue remembers that 'when the apples were coming down, and people had apples to give you [...] we would cook 'em up, because my husband loved apple pie and things like that, and I would freeze all that then in bags'.[12] Mary Berry's book concludes with a range of recipes, all of which include instructions on how to freeze, thaw and serve the dish. Artichoke soup, for example, perfect for the winter-time, could once cooled be covered, labelled and frozen, and then slowly thawed over a low heat on the hob when it was required. Recipes like 'Farmhouse Pâté' and 'Melon Cocktail', meanwhile, would need to be defrosted in a refrigerator overnight, while 'Bacon and Onion Quiche', frozen pizzas, meat pies and 'Boeuf Bourguignonne' could all be reheated in the oven straight from the freezer. A whole host of fresh, colourful ice cream dishes from 'Iced Lemon Soufflés' to iced syllabub cream and sorbets could now be served at home, to the delight of Mary Berry's then young children.[13]

Although fridges and freezers could reduce waste, they also led to changes in consumer behaviour that increased it. In the 1950s and 60s, meals were carefully planned in advance (roast dinner on a Sunday, leftovers fashioned into bubble and squeak on a Monday, bread and butter with tins of corned beef for a mid-week tea, fish and chips on a Friday, perhaps...). Though of course meal planning continued after these decades for many, fridges and freezers meant we could stock up on food that we might not end up using before it went off. In more recent years, the sociologist David Evans has described fridges and freezers – somewhat morbidly – as 'coffins of decay'.[14] Who among us hasn't pulled out a mouldy tomato or two from the back of the fridge, or forgotten a ready meal in the freezer for years, long past its use-by date?

These new household gadgets, along with televisions and cars, were emblematic of the rise of consumer wealth in Britain as the economy recovered after the war. In disposable income, by the beginning of the 1960s households had about half as much again as they had had in the 1950s.[15] Ordinary people had access to more food than ever before, and between 1961 and 1992 it got cheaper, falling by 47 per cent in price.[16] As food historian Lizzie Collingham states, 'from some point in the mid-1950s virtually the entire population in developed countries could afford to eat as much as they wanted'.[17] Unsurprisingly, then, in the three

Figure 32. A woman taking jars of food from a Hotpoint deluxe refrigerator, taken by British Photographic Advertising Limited in 1950.

decades immediately following the Second World War, the population increased by 5.9 million people.[18] These 'baby boomers'* were born into a healing, expanding, plentiful, optimistic and youthful Britain. In 1960, around 40 per cent of the population was under the age of twenty-five.[19]

As well as following America's lead in embracing the fridge, the UK soon adopted another American innovation: the supermarket. The first supermarket, named King Kullen, opened in the Queens borough of New York on 4 August 1930, using the slogan 'Pile it high. Sell it low'. The UK followed suit in 1951 when a Premier Supermarket was established in Streatham in South London.† Iceland, selling exclusively frozen food, opened in 1970 in Oswestry, Shropshire. Though today it is hard to imagine, given our dependence on self-service supermarkets,‡ customers were universally served their groceries directly by shop assistants until the mid-twentieth century. Daily trips to various local shops were soon replaced with a bigger weekly shop, as supermarkets expanded from a total of 50 in 1950 to 3,400 by 1969.[20] With self-service supermarket shelves, packaging became more

---

* Those born between 1946 and 1964.
† The claim to being the first ever supermarket is hotly contested, and depends on how you define a supermarket. The first self-service store, named Piggly Wiggly, was opened in 1916 by Clarence Saunders in Memphis, Tennessee, and the brand quickly franchised. In the UK, the self-service concept was adopted in the 1940s; the Co-operative Society opened the first fully self-service British grocery shop in January 1948, in Manor Park, London. It is worth noting, too, that Victorian consumerism had brought about the establishment of several grocery shop chains. Sainsbury's opened its first branch in 1869 in London, and by 1919 had more than 123 branches. See Kim Humphery, *Shelf Life: Supermarkets and the Changing Cultures of Consumption* (Cambridge, 1998), p. 31; Rachel Black, 'Supermarkets', in Carl A. Zimring and William L. Rathje (eds.), *Encyclopedia of Consumption and Waste: The Social Science of Garbage* (Los Angeles, CA, 2012), pp. 879–81.
‡ In 2020, less than 2 per cent of shoppers bought their food in independent stores, preferring instead the convenience and cheap prices of supermarkets. Lindsay Miles, *The Less Waste No Fuss Kitchen: Simple Steps to Shop, Cook and Eat Sustainably* (London, 2020), p. 8.

than simply a container to transport or store food. In aisles and aisles of products with different brands side by side, the packaging became an advert for its contents, having to compete to catch the attention of the busy shopper. One result was the arrival of bright sugary cereal packets, like Frosties (1954), Sugar Puffs (1957) and Coco Pops (1960), all of which were marketed towards children with special characters – Tony the Tiger (who started out advertising Kellogg's Sugar Ricicles but was replaced in 1960 by Noddy and moved to Frosties), Jeremy the Bear (the original mascot for Sugar Puffs before he was replaced with the Honey Monster in the 1970s), and Coco the Monkey.

Rising consumer culture also combined with the food sciences to create fast food, a concept with its roots in 1920s America but which took off from 1948, when Dick and Maurice McDonald transformed their successful drive-in restaurant in San Bernardino California by standardising items and assigning specific tasks to each worker. It wasn't just McDonald's that was born in the post-war era. Burger King began franchising within the US in 1961, Glen Bell established Taco Bell in 1962, and Kentucky Fried Chicken (KFC) arrived in Britain in 1964.

We can count supermarkets, household fridges and freezers and fast food as part of the so-called 'throw-away society' that was born in the last half of the twentieth century. Food waste doubled between 1939 and 1976.[21] Household waste as a whole, including new types of food packaging, had also essentially doubled by the late 1960s compared to pre-war levels.[22] What a change over a few decades! If during the war years, wasting food threatened lives and nations, from the 1950s onwards waste from the food industry became an all-too-recognisable symptom of surplus, capitalist growth and mass-consumerism fuelled by a revolutionised, hyper-productive agricultural system and a globalised food chain. In this chapter, we'll trace the activities of some counter-cultural movements that eventually led governments to take action on food waste issues.

The New Environmentalist Movement

Spring is my favourite season. When the sound of bird song wakes us from sleep, insects flutter in the sunshine as they explore and pollinate, flowers begin to unfold, new, hopeful plants emerge overnight from the darkness of the soil, and life blossoms once again after the cold of winter. Imagine instead a spring clouded in 'a strange stillness', with withered crops, unhatched chicks, empty streams and silence in place of the joyful morning chorus: 'a spring without voices'. This was the picture of a not-too-distant-future that the American ecologist Rachel Carson painted in her influential book *Silent Spring* of 1962. The natural world was choking, she warned, on man-made chemical pollutants that contaminate earth and water, passing through the food chain of countless plants and animals until they infect our very bones.[23]

When Carson was writing, new pesticides were being produced at an alarming rate. After the initial boom in manufactured chemical pesticides of the industrial nineteenth century, and further work to synthesise chemicals early in the next century, more than 200 new varieties of chemical pesticide had appeared between the mid-1940s and the start of the 1960s.[24] The organic compound DDT (dichlorodiphenyltrichloroethane), whose effectiveness as a pesticide was discovered in 1939, was widely used during the later years of the Second World War to protect Allied Forces personnel serving in tropical regions from insect-borne diseases such as malaria and typhus. After the war, DDT was made available to farmers as a pesticide for use in agriculture, replacing other (plant-based) pesticides that we have already come across in this book like nicotine and pyrethrum (from chrysanthemums). As we've also already seen, before this point in time, farmers commonly ensured that pests did not get out of hand by rotating crops or leaving arable land fallow. New pesticides like DDT offered a quick way of maximising agricultural yields, but as they were sprayed indiscriminately over farms, forests and gardens, they also killed wildlife in the surrounding areas. Without sufficient research and in an age when capitalist growth reigned supreme, the short-term need to limit crop damage and waste from insects, rodents and other pests was prioritised over longer-term consequences of pesticide use. Historians of the future will surely wonder, Carson wrote, how seemingly 'intelligent beings seek to control a few unwanted species by a method that contaminated the entire environment and brought the threat of disease and death even to their own kind'. Carson warned that we were blithely infecting the entire food system with harmful substances that could threaten our genetics, health and survival. Even more scandalously, the losses caused by natural pests were not harming food supplies on a national scale, only the profits of producers. Huge surplus stores of crops were costing the American taxpayer more than one billion dollars a year to store.

Carson's book illuminated the delicate balance of the natural world, and the counter-productive knock-on effects of pesticide use. New evidence suggested, for example, that pesticides reduced nitrogen-fixing bacteria in soil, without which it was starved of key nutrients that make our crops grow. Plants branded as 'weeds' and ruthlessly destroyed by chemicals may also have been performing the important job of adding nutrients to the soil, ultimately helping us to produce our food.[25]

Although *Silent Spring* focused on the United States, Carson's claims certainly resonated in Britain. A mysterious 'fox death' was first reported in November 1959 from Northamptonshire. Soon cases were reported across the country, and it was estimated that 1,300 foxes died overall after displaying an alarming array of symptoms including confusedly wandering in circles and falling into violent convulsions. Then, birds began to drop dead from the sky in droves. At the royal estates at Sandringham in Norfolk, pheasants, partridges, wood pigeons, greenfinches and sparrowhawks were among the 142 birds found dead in just an 11½-hour period. In spring 1961 a special committee of the House of Commons was set up to find out the cause of nature's mysterious malady. The answer lay again in toxic pesticides, including dieldrin, aldrin and heptachlor, which were routinely added to seeds to kill off insects that threatened crops. Birds were the unwitting victims of attempts to lessen agricultural wastage for human consumption, and foxes too had in turn been poisoned, probably from eating infected birds or rodents.[26]

A huge bestseller, Carson's revolutionary book led to the eventual banning of the particularly toxic pesticide DDT, but more broadly kick-started the so-called 'new environmentalist' movement in the United States and across the Western world. Among its chief targets was the worldwide spread of an industrialised agricultural system that sought to artificially increase yields and reduce the losses of profitable crops in the face of mounting global population levels. Since the 1950s the world – driven by the West – had embarked on the so-called Green Revolution, which introduced new pesticides, as well as mechanisation, and artificial fertilisers to replace natural dung, food waste and crop rotation. In Britain, post-war agricultural improvements of this kind resulted in wheat yields rising from 19.1 hundredweight (cwt)* per acre to 28.2 cwt in 1968/9, with barley similarly rising from 17.8 to 27.4 cwt per acre. Moreover, Robert Bakewell's eighteenth-century principles of selective breeding were cranked up a gear with the creation of new, more productive, breeds of livestock, meaning that dairy cows could produce 815 gallons of milk on average instead of 544 per year, and new poultry hybrids could produce more eggs, from 108 a year in the late 1940s to 211 per hen in 1969.[27]

Most significantly, this focus on selective breeding led to new high-yield dwarf cereal varieties being developed, which with a shorter stem could support the weight of more grain without falling over and wasting back into the ground. A new type of

---

* An imperial unit of weight equivalent to 112 pounds or 50.8kg.

wheat, growing about half the height of natural varieties, had been developed by Japanese agronomists in 1925 based on the genetic calculations carried out by the Austrian monk Gregor Mendel in the nineteenth century. Called Norin 10, after the Second World War it caught the attention of an agronomist in the US army who transported it back to America and then to Mexico, where Norman Borlaug, a recent doctoral graduate, got to work combining it with local varieties with the backing of the US government and the Rockefeller Foundation. From Mexico, the Green Revolution spread across the world, transforming the infrastructure of many poorer countries. A new 'miracle rice' variety, IR8, was developed in the Philippines by the International Rice Research Institute under the sponsorship of the Rockefeller Foundation. After experiments encompassing 10,000 different rice varieties, IR8 was eventually formed by crossbreeding an Indonesian strain and a dwarf variety from China. Like Norin 10, it produced much higher yields and was more disease resistant;

Figure 33. Seeds of the new rice variety IR8 being packed to distribute to farmers in the Philippines in the 1960s.

IR8 generated 10,000 pounds per acre compared to the 1,330 pounds per acre of the local rice, an almost ten-fold increase.[28] As these new varieties of crops spread, in certain parts of the world food supplies dramatically increased and disastrous famines were averted. Without these new productive crops, by 1996 the world would have had to find new farmland equivalent in size to the whole of North America in order to sustain the growing population.[29] Norman Borlaug, accordingly, won the Nobel Peace Prize in 1970.

These cutting-edge crops, however, relied on the newly developed pesticides and artificial fertilisers to be effective. In Britain, *The Ecologist* magazine was established in 1970. Its authors wrote regularly about the worrying environmental impact of fertilisers and pesticides, and were often critical of the Green Revolution. Michael Allaby, for instance, wrote that IR8 needs 70 to 90 lb of fertiliser per acre, which then runs into streams, causing algae to grow in the stew of nutrients, ultimately poisoning fish and making the water unusable.[30] As these novel varieties overtook local species across the world, a reliance on monocultures also meant a reduction of biodiversity that was ultimately counterproductive, since it left our food system more susceptible to disease and crop loss.[31] Even if the new crops were initially disease resistant, as pesticide, herbicide and fungicide use increased exponentially, insects and viruses evolved immunity to the chemicals, leading to more and stronger toxins having to be introduced to fight them. 'At least we will be feeding people, even if we are destroying their soils and poisoning them to do it,' Allaby concluded sardonically.[32]

In the UK, the New Environmentalist movement emerged within the context of growing fears of nuclear war and environmental disaster. In London's Great Smog of 1952 more than ten thousand people were killed by a lethal mixture of fog and smoke so thick that it seemed even a ray of sunshine could no longer reach the UK's capital; while in March 1967 a devastating oil spill from the tanker *Torrey Canyon*, which ran aground off the western tip of Cornwall, killed many thousands of seabirds. Global population growth was putting a visible strain on agricultural resources. What's more, in the 1960s and 1970s, scientific opinion was reaching a consensus that humans were having a detrimental impact on climate change, and this concern would lead, gradually, to a change of focus – to mitigating the impact of pollutants. The 1970s saw the establishment of several environmental groups in the UK, including Greenpeace (1971), Friends of the Earth UK (1971) and PEOPLE (1973), which would later become the Green Party. Environmental consciousness was gaining followers, throwing a spotlight on the entire food system.

The reliance on controlled tipping to dispose of the country's swelling stores of rubbish was another key concern of the movement. During the 1970s and 80s public awareness and fears were growing that landfill sites were filling up and that the country was running out of space for new ones. Ash and cinders made up an increasingly small proportion of domestic rubbish, as gas and electric took over from coal fires in the job of heating our homes, but bulky plastic, glass, paper and other consumer containers exploded in volume in the supermarket age. Refillable glass bottles – for fruit juice, milk and for soft and alcoholic drinks – were slowly phased out over the course of the late twentieth century, sent instead via rubbish bins to tips. The first public protest of Friends of the Earth UK, in 1971, involved dumping 1,500 Schweppes glass bottles outside the London HQ of Cadbury Schweppes, after the company announced that it was moving to disposable bottles. Friends of the Earth hoped the protest would 'promote re-use and better use of the planet's resources', even though it was ultimately unsuccessful in changing the company's mind.[33] By the 1960s, plastic packaging in particular was becoming a problem. Synthetic plastics had been invented at the start of the twentieth century, boomed during the war years, and their use in the food industry became commonplace from the 1950s. Plastic

Figure 34. Friends of the Earth protesting against Cadbury Schweppes for removing their refillable bottles in 1971, raising awareness about the extra waste that would end up in landfill.

debris was soon found in the oceans, and some raised concerns about the environmental impacts of these unnatural substances. The rise of plastics continued, however; by 1992 nearly 10 per cent of all household waste was made up of plastic containers.[34]

The disposal of leftover food into landfills also harms the environment. Compressed and deprived of oxygen, food waste produces huge amounts of methane, a gas which is 25 per cent more damaging to the Earth's atmosphere than carbon dioxide.[35] The impassioned wartime drive to collect and reuse leftovers as pig food fell away with relative ease in peacetime, and soon food waste was commonly dumped with other domestic rubbish in landfill sites. Domestic pig-keeping declined drastically, and even those urban areas like Birmingham and Manchester whose councils continued to find value in collecting food waste for recycling only did so into the early 1960s.[36] Jean Liedloff, the New York author who became a founding member of *The Ecologist* in London, complained that a report from 'The Countryside in 1970' conference under its president, the Duke of Edinburgh, had failed to take into account the 'long term advantages of such techniques as recycling, especially of organic wastes for agricultural purposes as an alternative to artificial fertilisers'.[37]

As we've seen, things thrown into landfill sites – including food waste and packaging – have long been reclaimed, resold and reused. The term 'recycling', however, emerged for the first time in the New Environmentalist movement. Its meaning was distinct from the 'salvaging' that had taken place before in that it was directed towards saving the environment rather than simply making the most money or use out of an item. This was also not the first time that the public had become concerned about tipping. Late Victorian fears, however, had centred on the poor environmental conditions – the smells, fumes and diseases – that these sites could breed, rather than on their wider impact on the natural environment for its own sake.

Recycling as an alternative to tipping was at first counter-cultural and change was gradual. The UK's first bottle bank opened in 1977, and London's first distinct mixed recycling centre was established in 1985. The European Union, which the UK had joined in 1973, also began to take note, and in 1999 the Landfill Directive set targets for its member states to reduce the amount of biodegradable municipal waste being sent to landfill. Despite mounting pressure, the UK government did not enforce recycling, and in 1996 the country was still sending 90 per cent of all waste to landfill, with only 5 per cent being recycled.[38] Plastic was – and is – notoriously difficult to recycle as it comes in numerous different forms that are time consuming to separate. Still, that same year the UK government took its first steps to clamp down on landfill, with the first green tax on its use. In 1993 Kensington and Chelsea in London had become the first local authority to offer household recycling. But it wouldn't be until 2010, as laid out in a 2003 law, that the government in England made it mandatory for all local authorities to provide kerbside recycling of at least two types of materials.

McSpotlight

As the New Environmentalist movement was beginning to catch on in the 1970s, American fast food chains were establishing themselves in Britain. Wimpy, which served hamburgers and milkshakes, arrived in the 1950s, providing many in Britain with their first taste of fast food other than perhaps a traditional fish and chips on a Friday night (notably, a relic of the religious fasting laws laid out in Chapter 1). When McDonald's was first established here in 1974, it soon dominated the fast food market in the UK. These fun, colourful restaurants were the favourite haunt of many young people, who experienced the novelty of American burgers, milkshakes and sodas for the first time, often with the disapproval of their more conservative parents. According to the sociologist Annechen Bahr Bugge, American fast food chains became 'emblematic' of the emergent youth culture, which was, perhaps for the first time, clearly distinct from 'adults and "adult culture".[39] McDonald's, and other American fast food brands, were also an important site of youth employment. Owen, for instance, who grew up in London's Westminster and Richmond boroughs in the late 1980s, told me that he began working in the American burger chain Wendy's (or 'Wendy' as it was then called) at fifteen years old. He recalled the familiar scene of '200 to 300' young people trying to get served in the pubs on Friday and Saturday nights before piling into the restaurant. After a shift, 'your shoes would slide slightly until the pavement cleaned them', he remembers.[40]

Before long, however, fast-food chains, and especially McDonald's, were subject to increasing public criticism on a variety of fronts. Most significant was the so-called 'McLibel' trial which began in 1990 when McDonald's sued members of London Greenpeace, a small independent anarchist organisation formed in 1972 (and unrelated to Greenpeace International). Members had distributed a six-page leaflet titled 'What's wrong with McDonald's: everything they don't want you to know' over the last four years of the 1980s. The leaflet accused the company of myriad sins including cruelty to animals, worker exploitation, promoting an unhealthy diet (especially to children), and environmental crimes. Moreover, the activists were largely vegetarian, and they claimed that in encouraging meat-eating, McDonald's was responsible for wasting grain and soy that would better go to feeding the world's expanding population.[41] During the case, spurred on by the advent of the internet in the early 1990s, a new website called 'McSpotlight' was able to translate the newspaper into different languages and spread information criticising McDonald's around the world.

In English libel law the defendants had to prove the truth of their statements. Helen Steel and David Morris' attempts to do so would turn the lawsuit into the longest-running libel case in English history. After the two defendants were denied legal aid, Steel and Morris were assisted on a *pro bono* basis by a young barrister named Keir Starmer (who would later become leader of the Labour party). Although when the case finally came to an end in 1997 McDonald's was awarded £60,000 by the courts (which it did not collect, and which was later reduced to £40,000), in examining the accusations made against the company, many of its shortcomings were exposed on a national and global scale, causing huge damage in the court of public opinion.* As Bob Langert, leader of McDonald's sustainability efforts from the late 1980s until 2015, described, its image was 'suddenly warped from

---

* The European Court of Human Rights ruled later that the two defendants, Helen Steel and David Morris, had been subject to an unfair trial and the UK government was ordered to pay them £57,000.

a symbol of happiness and fun to an icon of waste amid a disposable society'.[42] Georgina, growing up in the mid-1990s in Bristol remembers that there were posters at the bus stop outside her school protesting against McDonald's. 'My friends and I never went to McDonald's,' she says, 'we thought they were evil – I was a vegetarian anyway so [there] wasn't much point, we objected on grounds of principle to fast food.'[43]

It wasn't just McDonald's itself that came under fire but their suppliers, those who fed and slaughtered the animals that would end up in Big Macs and Happy Meals. One of the major issues exposed was the conditions that McDonald's suppliers kept their animals in. Images of sentient creatures treated like cogs in a machine, caged and cooped up for their short lives, before being killed in inhumane ways, were shared in the documentary film *McLibel* of 2005.[*] Helen Steel described how in one farm, around 1,000 chicks deemed useless were simply gassed to death, their life and the food they could have produced ruthlessly wasted in the name of higher profits.[44] The factory farming system, indeed, works on efficiency and creating the least amount of waste possible, but only when 'waste' is defined as *profitable* meat or agricultural outputs. Anything else is rejected from the system.

The BSE, or mad cow disease, outbreak in Britain in the 1980s and 90s also struck a major blow to McDonald's' public image. Bovine Spongiform Encephalopathy is a neurological disease of cattle which, when it spreads to humans, results in vCJD (Variant Creutzfeldt-Jakob disease). BSE-infected meat killed the first human in 1995. The outbreak was caused by feeding livestock rendered meat-and-bone meal (MBM), the ground-up remains of animals. This is commonly practised as a way of using up all of an animal after its slaughter and to provide extra protein to

---

[*] The film had originally appeared after the trial in 1997 but was expanded and released in a longer version in 2005.

increase the milk yields of dairy cows.* For many, the BSE crisis epitomised the criticisms levied at the capitalist food chain, which for the sake of profit had seemingly corrupted nature by feeding animal remains to herbivores. Similarly, in the late 1960s and 1970s, the new practice of recycling chicken protein into animal feed allowed salmonella to spread.[45] While environmentalists celebrate recycling, these commercial practices took the concept to an aberrant extreme.

Concerns over landfill raised by environmentalists in the 1970s had reached the mainstream in the 1980s and McDonald's once again became the target of public outrage. In particular, its PSF, or polystyrene foam, clamshell, which housed McDonald's' burgers, had come to symbolise the evils of a wasteful global food system. This newly invented container was chosen as an economical way of protecting the food and to keep it from getting cold and spoiling, but it was sent to landfill, where it would never fully break down. In response, McDonald's acted to try to develop a recycling programme, with a separate bin for PSF in their restaurants. But customers who had come to McDonald's

---

* In response to the BSE outbreak, a 1988 UK law banned ruminant protein being fed to ruminants, and since 1994, the EU has banned the use of processed animal protein (PAP) from mammals being used in the feed of ruminants. In 2001 this ban on consuming PAP was controversially extended to all farm animals in the UK (and across the EU in 2002), ending the practice of feeding food waste to pigs. This was a response to the foot-and-mouth outbreak in the UK in 2001, which came from a farm feeding untreated swill to pigs. Unlike herbivorous cows, pigs and poultry are omnivores and so there is nothing inherently unhealthy about feeding them PAP from other animals once it has been heated to kill pathogens. EU legislation from August 2021 allows PAP from pigs to be used in poultry feed and likewise PAP from poultry to be used in pig feed. This move is presented as a way of reducing the need to grow animal grains, especially soya, the farming of which in the Amazon basin has huge environmental consequences, as well as providing the economic benefit to farmers in order to compete outside of the EU. Bans remain on the use of PAP to feed cows and sheep and on 'intra species recycling', when animals are fed remnants of animals from the same species. As of 2023, the UK has yet to follow suit.

for fast – and thought-free – food ended up dumping all their trash into the garbage bin, rather than taking the time to separate it out. Any PSF that was recycled became contaminated with food waste, which quickly reeked and attracted vermin at the recycling plant. By the end of 1990, McDonald's chose to phase out the PSF packaging and opt for a cardboard version instead. Still, a major issue was the leftover food itself, which accounted in fact for 90 per cent of McDonald's' waste. Pilot tests successfully created useable soil from 'fry cartons, wraps, and food' through composting. According to the then head of McDonald's' sustainability drive, Bob Langert, food composting in this way is still far from commonplace in restaurants, but he believes it to be the way forward for the company. In fact, it is worth noting that in response to heavy criticism, McDonald's led the way in addressing issues with waste in the food industry. Langert argues that before the 1990s, the concept of a sustainable food chain was basically unknown in the food industry, retailers believing that

Figure 35. McDonald's clamshell containers, which were a source of environmental controversy in the 1980s. Dated between 1975 and 1990.

they could not be responsible for verifying the integrity or sustainability of goods sourced from overseas.[46] Still, this spotlight on the fast-food giant and the environmental problems associated with the modern food industry – set up to feed a bursting, increasingly urban, consumerist society – were exposed for all to see, if they weren't too squeamish to look.

Radical Reclamation

It is a radical political act in today's wasteful society to recover large amounts of food in an organized and consistent manner to share with the hungry.

Keith McHenry, *Hungry for Peace* (2012).

Also emerging in the 1970s and also originating in the US, punk music was deeply tied to the environmental, anti-globalisation and anti-establishment movements of the era. By the 1980s, in the context of Margaret Thatcher's Conservative government, and following years of inflation, recession and unemployment in the 1970s, the more directly political anarcho-punk movement had arisen in Britain. Fuelled by political rage and youthful rebellion, half-screamed lyrics railed against social inequalities and poverty, the corruption of powerful elites, unjust wars (the Falklands War, in particular, as well as the looming overarching threat of nuclear war), animal cruelty and very often against meat-eating more generally. Formed in 1977 in Epping Forest, Crass were at the heart of anarcho-punk. In 1983, they released the longest punk song ever written in the form of a one-song album entitled *Yes Sir, I Will*. Written by drummer Penny Rimbaud, the lyrics were a kind of political manifesto exposing the corruption of the establishment over forty-three minutes, performed over the top of an improvised music track. Around halfway through the rampage, two lines shamed governments for destroying food rather than giving it to those in need. Reflecting the views of many punks of the era, the piece sees food waste as the deliberate injustice of an economic-political system in which access to resources like food are unevenly distributed between the rich and poor. British punk in the 1980s was generally antagonistic towards America, blaming it for the rise of mass consumerism and the waste that came with it: 'Smash the Mac, smash the Big Mac', Crass sang in 1984, ending notoriously with the formidable couplet: 'E.T. go home [...] Mickey Mouse fuck off'. As Dylan Clark has argued, 'the waste of food and the protection of waste were seen by punks as the avaricious gluttony of American society'.[47] As well as squatting in shared premises in defiance of the unequal distribution of wealth, some punks were known for 'dumpster diving', foraging for food in bins behind shops and supermarkets as a means of circumnavigating the capitalist food system, which perceptibly left perfectly good food to rot while people went hungry.

You might have heard this practice today being called 'freeganism'. The term was first coined by one of the founders of Food Not Bombs, Keith McHenry, in the mid-1990s.* Food Not Bombs is another counter-cultural food recovery initiative, founded in the 1980s in the United States, which quickly went global. Working in a grocery store in Cambridge, Massachusetts, McHenry remembered a 'dilapidated public housing building' just a few blocks from the shop, around which undernourished children and their mothers 'huddled in the cold'. Right next door was a modern tower block 'where scientists were busy designing guidance systems for intercontinental nuclear missiles' to feed the Cold War arms race. This visible injustice made it abundantly clear to the young activist that people needed food, not bombs. The anarchist

---

* The other founders are Mira Brown, C. T. Lawrence Butler, Brian Feigenbaum, Susan Eaton, Amy Rothstein, Jessie Constable and Jo Swanson.

group intended to 'promote life, not death', driven by the desire to redirect the huge financial resources put into the US military towards fighting poverty and hunger within the country.[48]

One of the major ways it did so was by redistributing leftovers from the food industry, like those discarded daily by the grocery store McHenry had worked at. It wasn't hard to find unsold bread at bakeries at the end of a day, or surplus perishable fruit and vegetables from co-op stores. In 1991, the sociologist Aileen O'Carroll wrote in an edition of the anarchist paper *Workers Solidarity* that despite the fact that 30 million people face starvation, 'beef, butter and wine mountains rot in European warehouses', and 'farmers are ploughing crops back into the land'. In the EU, indeed, the Common Agricultural Policy that was first introduced in 1962, giving farmers in the member states subsidies to keep prices stabilised and supplies plentiful, had led to huge levels of overproduction. Excess stocks in the form of so-called 'beef mountains' or 'milk lakes' were sold off cheaply by the EU. This, O'Carroll – and later Keith McHenry – used as evidence for the failings of capitalism, demonstrating that hunger is created by political systems in which food is wasted and resources unfairly distributed.[49] By reclaiming lost food, Food Not Bombs sought to non-violently resist the state and the hierarchical structure it was built on. The founders saw this as a mission with deep historical roots: from the radical English Diggers of 1649 who took over and cultivated common land on St George's Hill near Weybridge in Surrey in order to grow food and share it among those who worked the land; to the San Francisco Diggers of the late 1960s, who adopted the same name and dished out leftovers from markets and dumpsters every day to homeless people.

In their first book, *Food Not Bombs: How to Feed the Hungry and Build Community* (1992), the founders of Food Not Bombs simply stated: 'The world produces enough food to feed everyone, if distributed equally. There is an abundance of food. In fact, in this country, every day in every city, far more edible food is *discarded* than is needed to feed those who do not have enough to eat.[50] The pamphlet included instructions on how to set up local 'chapters', which were to be run as autonomous, democratic and nonviolent groups. In looking for surplus food to redistribute to those in need, members were encouraged to approach shops – local co-ops and health food stores being the best bet – and 'ask the workers at these businesses if they have any edible food that they regularly throw away, and, if so, that they would be willing to give to you'. Be sure to mention, it notes, that giving the surplus food away will save the business money on their waste disposal costs, which 'keep growing each year as more and more landfills become exhausted'. Since these businesses by design overstocked perishable foodstuffs so that customers would not be disappointed, Food Not Bombs members could expect to see a continuous stream of food waste. Next, they were to set up regular collections to transfer the food to local soup kitchens and shelters. Once this network was established, members could take a percentage of the food collected to cook publicly and give out to homeless people. This was an opportunity to make a difference to people's lives but also to make the issues of homelessness and the Food Not Bombs message visible.[51]

While most food was to be recovered from waste or donations, the leaders recognised the need to buy certain ingredients, including cooking oils, spices and some dry goods. Anything that was bought, though, should be taken from local co-operatives or health food stores so that it was organically grown and wrapped in less packaging.[52] The book contains recipes – notably all vegan – for Food Not Bombs members who gave out food to the hungry, or needed to feed large numbers of people at political demonstrations and occupations. Most simple of all was the day-old or surplus bread and pastries that were collected from local bakeries: 'place the cut loaves in a large clear plastic container with lid. Attach a set of tongs to the container with wire or string.' Next were raw vegetables, which were 'greatly appreciated by people on the streets'. They could simply be washed and mixed into a salad, which was to be dressed only immediately before serving since 'a dressed salad will not stay fresh overnight; it becomes soggy and unappetizing'. Recommended breakfast dishes included oatmeal, granola and scrambled tofu. The latter also featured in lunch recipes such as 'Tofu Spinach Lasagna' and 'Tofu Sandwich Spread', both of which were intended to serve 100 people, which gives a sense of the scale of events that Food Not Bombs were running:

TOFU SANDWICH SPREAD

Makes: 100 sandwiches
Need: medium mixing bowl, very large mixing bowl
Prep time: 2 hours

3 cups miso
3 cups water
8 cups tahini
25 lb crumbled tofu
25 lemons, juice of

Optional:
½ cup garlic powder
8 cups diced onion
8 cups diced celery

In a bowl, mix the miso and water into a smooth paste. Add tahini and mix (add additional water to make a smooth, creamy paste). Drain tofu of excess water and crumble by hand into a very large bowl. Squeeze the juice of the lemons over the tofu. Add miso/tahini mixture and mix well. Add optional ingredients. Spread on your favorite bread with lettuce, sprouts, and tomato slices.[53]

According to the founder McHenry, this recipe came about after a tofu company didn't know what to do with the scraps that had broken off tofu cubes, delivering 'two or three five gallon buckets' filled with bits of tofu for the group to make use of.[54]

McHenry explained how at the end of a food distribution rally, any food left over was to be donated to another community project or, if not, divided among the volunteers. Storing it for a future event would likely reduce its nutritional value and mean that it was more likely to spoil, especially given the continual flow of surplus being collected from the food industry each day.[55] Members, too, were encouraged to grow food in community gardens, as a way of avoiding the produce of the global food industry and the environmental damage caused by pesticide and fertiliser use. Composting kitchen waste was recommended in much the same way – alternate layers of 'greens' and 'browns' – as the Ministry of Food had instructed in the Second World War (see Chapter 5). A good practice, McHenry wrote, was to occasionally poke holes roughly 6 inches apart through the whole compost pile to help aerate the mixture and encourage the growth of aerobic bacteria that form the compost out of the waste. The compost should also stay wet so might require watering if you didn't live in a rainy area.[56]

In the United States, the food activities of Food Not Bombs in the 1980s and beyond were sometimes met with the resistance of the authorities. Like the activities of 'dumpster divers', the practice of giving out free food could be viewed as an affront to established society and economic norms. The books or manuals published by the founders spend a great deal of time elaborating on how to resist and respond to arrests or other rough treatment from the police. In the founders' opinion, the 100 arrests made in San Francisco in 1988 marked the beginning of a more aggressive corporate and governmental response to the group's actions, leading to further 'arrests, beatings, disinformation and litigation in an attempt to silence'. That summer, in San Francisco,

California, many volunteers were arrested for the 'political state-ment' inherent in sharing free vegan meals with the hungry of the city without permission.[57]

Making it Mainstream

The leading food waste activist Tristram Stuart has been collecting food waste since about the age of fifteen. In the 'hey-day' of his freeganism, Stuart lived out of bins in north London, routinely finding ten full sacks of food – about a fifth of a tonne of waste – thrown in the bins of his local organic fruit and vegetable shop each Friday evening.[58] As we've explored, dumpster diving or 'skip digging' was already being practised by fringe anti-establishment or environmental activist groups of the late twentieth century. Stuart tells me he remembers that the crowds camping at the Newbury Bypass protest in 1996, who were campaigning to prevent the clearance of hundreds of acres of woodland for the construction of the new road, were fed on food that had been discarded. What Stuart did, however, was to bring the idea of freeganism and an awareness of commercial food waste to the wider public. At a time when the government was spending very little money on food waste issues, in 2002 Stuart initiated a media storm when he revealed the shocking amount of good food being thrown into bins behind supermarkets. Although supermarkets do now on the whole donate more food than they did twenty years ago, the discarding of still-edible food items is still commonplace. Excitedly, a friend reported to me that she'd found a modern 'garden of Eden' behind an M&S in Barnes, a dumpster stocked with dozens of cartons of eggs, multiple types of fish that had just reached their use-by date, sweet potato fries, tuna pasta bake and other ready meals, berries and even a crate of wine (a bottle somewhere had smashed, leaving an offensive stain on some of the labels). In a chain of influential television appearances, Stuart was able to 'articulate this act of reclaiming food', dumpster diving, 'as a fundamental critique, not just of food waste, but of the entire paradigm of overproduction within the industry'.[59] The major problem, he argued, is that it is more attractive for shops – and especially supermarkets – to order in a surplus of food rather than to risk disappointing their customers with empty shelves. Because of the difference between cost and retail prices, it is cheaper to throw away multiple items rather than risk losing a single sale from a customer enquiring about an item that had become out of stock. Customers also expect the comforting sight of abundance during their shopping trips, which means full, often overstocked, shelves.[60]

Just as in Tudor times when gleaning was seen by some as incompatible with the landowner's rights to the fruits of their land, dumpster diving was (and often still is) seen as a threat to our modern conceptions of property ownership. Even though they were throwing away their produce, shops commonly discouraged the foraging of leftover food from their bins by opening otherwise untouched food packets (and thereby contaminating them), locking bins, or even poisoning food by spraying it with chemicals. These practices were possibly undertaken to adhere to UK law, since the Environmental Protection Act 1990 Waste Management Code of Practice maintains that all waste holders 'must act to keep waste safe against scavenging of waste by vandals, thieves, children, trespassers or animals'.[61] Dumpster diving falls under the terms of the 1968 Theft Act, meaning that those who practise it could in theory be prosecuted for trespassing, and in 2013 three men were arrested under the 1824 Vagrancy Act for taking tomatoes, mushrooms, cheese and Mr Kipling cakes from the bins behind a branch of the frozen-food store Iceland (though the case against them was later dropped).[62] Yet, as Tristram Stuart exposed, the main concern of food businesses is that customers might stop buying their produce if they realised it could be accessed for free from bins at the end of the day. Referring to London branches of Eat, a chain of sandwich shops that operated between 1996 and 2020, Stuart made the argument that after fifteen years of 'dumpster diving', he had met no one who could afford to shop in Eat. Gleaners were instead most often homeless people, elderly people having to supplement their pensions, or immigrants working for inadequate wages. In Islington he regularly met a woman whose glasses were held together with sticky tape, who would repeat the biblical instructions given by Jesus at the feeding of the five thousand: 'Gather the pieces that are left over. Let nothing be wasted.'[63]

Spurred on by such injustice, on 16 December 2009 Stuart launched an international movement when he hosted the first 'Feeding the 5000' event in London's Trafalgar Square. With the collaboration of FareShare, the UK's leading food distribution charity, which was first established in 1994 to feed homeless people and in 2004 extended its reach to assist all those in food poverty, Stuart sourced food that would otherwise have gone to waste, like the surplus on nearby farms. With this the team made a huge vegetable curry that fed 5000 people for free. The recipe for such included the following ingredients and quantities:

500kg potatoes
400kg rice
100kg tomatoes
20kg carrots
20kg lentils
20kg chickpeas
20kg butter
7kg salt

Other items like tomato paste, peas, or cream could be added as pleased, though the recipe concludes that there should be, 'no onions, garlic, mushrooms, processed foods, eggs, meat'.[64]

Again, the event garnered huge media attention, and with public awareness of food waste issues mounting, had the added effect of pressurising the major supermarkets to sign up with FareShare to donate their surplus stocks. Indeed, after a Feeding the 5000 event in Dublin, Tesco invested in FoodCloud, a business that connects retailers with local charities to donate food daily, running pilots at several Tesco stores in Ireland and then the UK.[65] This led to the revolutionary scheme FareShare Go, which notifies those local food charities and community groups who sign up of surplus being offered by nearby supermarkets for collection.

With Niki Charalampopoulou, Stuart went on to co-found Feedback, an international charity focused on food waste which aims to transform the food system into a more sustainable alternative through a huge variety of campaigns and activist programmes. In total, Feedback and its partners have hosted over fifty Feeding the 5000 events, based on rescued food alone, across the world.[66] Feedback's Gleaning Network* was then set up to rescue fruit and vegetables from farms, which are wasted for a number of reasons including overproduction so as not to disappoint contracts or in the light of particularly fruitful weather, changes or cancellations to supermarket orders, or to comply with supermarkets' cosmetic standards.[67] Tristram Stuart says that he regularly meets farmers who waste at least 20, and up to 40, per cent of their produce as a result.[68] Members of the Gleaning Network continue to volunteer to harvest any surplus from local farmers, before redistributing it to various projects and causes in their communities, like FareShare as well as smaller charities and community cooking days. Since 2012, Feedback's Gleaning Network has rescued over 500 tonnes of fruits and vegetables from going to waste.[69]

---

* Similar grass-root gleaning schemes exist in Europe, such as the Espigoladors in Spain, and RE-BON in France. From 2021 the Gleaning Network became more independent of Feedback, which continues to provide an extensive toolkit, along with advice and support for regional groups.

As public awareness of – and outrage at – the environmental and social issues surrounding food waste were increasing, the British government was forced to act with top-down initiatives. At the turn of the new millennium, the Department for Environment, Food & Rural Affairs (Defra) funded the non-profit Waste & Resources Action Programme (WRAP).* Focusing at first on packaging waste, WRAP soon turned to food waste as part of its drive to establish a circular economy in Britain. Based on the initial EU Waste Framework Directive of the 1970s, WRAP promoted the 'waste hierarchy' in which prevention of waste is prioritised over the creation of waste in the first place; ideally, waste should be recycled, or otherwise recovered (for example, by extracting energy from it), with disposal as the least preferred option.† In 2005, WRAP launched the Courtauld Commitment, which pressured the major British supermarkets to voluntarily agree to cutting their waste and publishing data on this topic. Between 2005 and 2009, the Courtauld Commitment helped block the wastage of 1.2 million tonnes of food and food packaging waste, which saved 3.3 million tonnes of CO_2e,‡ equivalent to the emissions of 500,000 round-the-world flights.[70] The scheme has recently been renewed for a fourth time into 2030, and focuses on delivering the UN's Sustainable Development Goal of a 50 per cent reduction in per capita food waste compared to the UK's 2007 baseline.

The world food price crisis that struck in 2007 and the first part of 2008 provided a catalyst for top-down action on food

---

* In 2014 WRAP left Defra and became an independent charity, but was still primarily funded by Defra, the Welsh government and Daera in Northern Ireland, with historical funding from the EU.

† The EU Waste Framework Directive 1975. Food waste was not explicitly considered a part of the EU Framework until the Directive (EU) 2018/851.

‡ This means carbon dioxide equivalent, and denotes the number of metric tons of CO_2 emissions with the same global warming potential as one metric ton of another greenhouse gas.

waste. Between 2005 and 2007, food prices had increased by a staggering 75 per cent.[71] Recurring periods of droughts, rises in the cost of oil, an increase in meat consumption (which depends on grains as animal feed) and the redirection of grain for use in the biofuels industry, were among the root causes of the disaster. Huge numbers of people across the world were pushed into food insecurity, a lack of consistent access to food and drink. In Pakistan, for example, an additional 17 million people between March 2007 and 2008 fell into this category, making a total number of 77 million hungry, approaching half the country's population at the time.[72] Riots broke out across parts of Asia and Africa, including Bangladesh, Burkina Faso, Senegal, Cameroon, Egypt and Morocco. Paired uncomfortably with the global financial crisis, the world food price crisis finally forced governments to move the matter of food waste to the centre of their agenda. WRAP's report from 2007, for example, found that only 2 per cent of food waste was being collected for composting or anaerobic digestion (where it is converted into useful biogas and fertiliser). It recognised the opportunity for the UK to 'take a lead in dealing with biowaste' and encouraged local authorities to introduce specific food waste collections.[73] In the same year, 2007, WRAP launched the popular Love Food Hate Waste campaign, which focused on educating the public on ways of cutting food waste in a series of public campaigns. Its website soon began to fill with food-saving tips and delicious recipes. An example from 2009 suggested using up broccoli stalks, which often go straight to the bin, and any leftover vegetables in a 'wholesome soup'. Bring a few potatoes, an onion, a couple of carrots, and a handful of pearl barley or lentils to a boil, simmering for ten minutes or so, it suggested, before adding the broccoli stalks. Whizzed up in a blender, all it then needed was a little crème fraîche or sour cream to swirl into the soup.[74]

Once harvested, life starts to decompose
In the eternal cycle from seed to soil.
And so we've mummified the nation's larder,
Embalmed it, so to speak, with chemical skills
Extending consumer's choice through every season,
The failures of nature redeemed by the genius of man.
[…]
Food can be kept for the day of the resurrection
To greet the saints with a savour of baked beans
But where, by then, will the tins have all been stacked?
Mountains of tins festooned with putrefied labels,
Mighty plateaux moulded from kitchen waste,
Shoals of tins floating in the seas,
Pesticide cans of indestructible poisons
Killing the fish and drifting back to the rivers […]

Robert Waller, 'Scabby Apples', *The Ecologist*, 1:3
(September 1970), p. 42.[75]

It was in the second half of the twentieth century that the issue of food waste took on the environmental and political gravitas that it commands today. This was a rapid and urgent departure from the way food waste had been understood in the past. As I've suggested, writers have long condemned wasting food as a needless dissipation of natural and financial resources, as well as a public health matter. It was only once the UK, along with the rest of the developed world, had committed to a globalised and industrialised food system, with mass consumerism and rapid economic growth at its core, that food waste – as well as new forms of food packaging – was identified as an issue that threatens the survival of the Earth itself. The poet and environmentalist Robert Waller, an associate editor of *The Ecologist* magazine, described in 1970 how in our quest to arrest the relentless cycle of decay, we had created a food system in which crops are smothered in dangerous chemicals, that ultimately infect waterways and foods with poisons, while the Earth is littered with non-degradable food packaging and festering mounds of kitchen waste.[76]

Despite activists' calls for change from the 1960s onwards, the development of the modern agricultural system and the consumerist society that fed from it forged ahead. A reliance on supermarkets as the source of our food, for example, was established by the end of the century. As butchers closed, more and more people bought their meat from supermarket shelves already processed and wrapped in plastic.[77] With the rise of household refrigeration and freezing, we drifted further away from nature and a closeness to the seasonality of foodstuffs. The days of Mrs Beeton's thrifty leftover menus, and of home-made bottled, salted and dried preserves, slipped away, to be replaced with takeaways and Tupperware ('Tupperware parties', where the nifty plastic packaging was sold, were a fad in the 1960s!). The mass consumption, and the waste that comes with it, symbolised by American fast food chains continued to expand, as McDonald's restaurants reached 1,000 in number in 1999.[78] UK household food waste of 6.7 million tons in 2008[79] looked pretty unforgiveable in the light of the global food crisis and the continued presence of poverty at home and abroad.

As we moved into the 2010s, the climate fears that had simmered fifty years previously began to bubble over. The Swedish teenager Greta Thunberg, born in 2003 into a world already enlightened to the catastrophic implications of global warming but seemingly unwilling to take the radical steps needed to curb it, ignited a global 'School strike for climate' movement (or 'Fridays for Future') in the summer of 2018 when she – at the age of just fifteen – began protesting. One straight-talking school student soon became the face of tabloid debates on the urgency of climate action. Questions concerning food waste are essential to this conversation: do our modern attempts at increasing agricultural yields with chemicals, new breeds and large-scale industrial farming cause more harm than good to the planet's future food supplies? How do supermarkets contribute to food waste on farms and in homes? How can we best reuse or dispose of food waste and food packaging in a more environmentally sustainable way? But answers, or at least action, would have to be put on hold, as the world's attention was soon to turn to another environmental disaster, which would alter our relationship to food waste once again.

7

FOOD WASTE IN THE TIME OF CORONAVIRUS

Lockdown Banana Bread

Ingredients
2 very ripe or overripe bananas
140g butter
140g caster sugar
140g self-raising flour
2 eggs
1 tsp baking powder
Pinch of salt

Method
Butter and line a loaf tin with baking paper, and preheat the oven to 175° C

Cream the butter and caster sugar together

Beat the eggs and add them to the butter and sugar

Mash the bananas with the back of a fork

Fold in the flour, with the baking powder, a pinch of salt, and the mashed bananas

Mix in any extra ingredients of your choosing (chocolate chips, nuts, dried fruit etc.)

Pour mixture into the baking tin and bake for around 50 minutes, checking regularly with a skewer or fork as the time it takes to cook can vary depending on what additional ingredients you used

> Let it cool before serving (and refrigerate or freeze it for future use).

Britons went a bit bonkers for banana bread during the first coronavirus lockdown, which began in March 2020, as many of us spent more time in the kitchen. The recipe is incredibly flexible, made up of flour, eggs, sugar and butter, with any overripe bananas people had left over. You could add a range of ingredients that might be found at the back of a cupboard, or that you'd overstocked in the initial stockpiling panic.

———•———

Over the last half century, supermarket shelves have heaved with the weight of more and more food: new ingredients, products and packaging. The carefully crafted illusion of infinite food supplies was quickly shattered in early 2020, however, when the threat of a new virus, Covid-19, became alarmingly apparent. Perhaps one of the enduring images of the coronavirus pandemic is that of empty supermarket shelves across the world, as consumers raced to stock up on essential long-lasting food supplies, especially pasta, rice, tinned vegetables and long-life milk (as well as – infamously – toilet paper). In the three weeks before the UK lockdown was announced on 23 March 2020 (coming into force on 26 March), an extra £1 billion worth of food was hoarded away in UK homes.[1] Our natural instinct to stockpile food resurfaced quickly, just as it had at the beginning of the First World War, more than 100 years before. Perhaps you remember the viral video shared on 19 March, which showed a critical care nurse breaking down exhausted and in tears after she found the supermarket shelves devoid of fruit and vegetables following a forty-eight-hour shift.[2] Years later, it's easy to forget how terrifying this initial period of relative scarcity was, as people feared they would not be able to get enough food to feed their families. For those with young children, with special diets or allergies, the situation was even more alarming, to say nothing of the real and daunting threat to health and life that the global pandemic brought to families and their loved ones.

Once the images of barren shelves had subsided, social media feeds were full of perfectly risen sourdough loaves. Easily rustled together with flour, eggs, sugar, butter and overly ripe bananas (to which whatever nuts or chocolate you had left in the cupboard could be added), banana bread is also a strong contender for the title of most snapped quarantine food. With more time on our hands, cooking and baking was a wholesome and rewarding pastime for those who could afford it, as well as a way of entertaining restless children being home-schooled. No wonder flour was regularly hard to come by throughout the first lockdown. (To meet the increased demand, a thousand-year-old mill, Sturminster Newton Mill in Dorset, which was first established in the eleventh century, opened up again to produce flour on a commercial scale.[3]) As supermarkets began placing limits on the amount of certain foods that customers could buy, including pasta, flour, UHT milk and some types of tinned vegetables, along with anti-bacterial wipes and toilet paper, they insisted these shortfalls were not caused by wider supply issues, but by hoarding.[*]

Yet, the pandemic was having an adverse effect across the food supply chain. At the same time that shoppers were finding bare shelves, British dairy farmers were having to throw away thousands of litres of perfectly good milk down the slurry pit. Contracts for milk destined specifically for hotels, airlines, restaurants and cafés were cancelled overnight as the hospitality industry shut its

---

[*] The major UK supermarkets rationed these items from early March 2020. Similar rules were reintroduced in September, as the 'second wave' of Covid took hold.

doors. Meanwhile, as countries began to close their borders in an effort to contain the virus, they also closed off these routes to seasonal foreign workers who picked crops during the harvest, in a further shock to food supply chains. In Germany farmers worried that their famed white asparagus crops would rot in the fields without harvesters from Eastern Europe, while the blueberries in Spain's Huelva province were no longer being picked by workers from Morocco.[4] In the UK, the government launched its Pick for Britain campaign, which hoped to recruit 70,000 people from the population to fill the gap in harvesters. From the well-groomed gardens of Clarence House in London, Prince Charles was called in to recruit Britons to sign up so as to make 'a vital contribution to the national effort'. Urging the public to readopt the wartime spirit (a powerful concept in the British imagination even for the majority of citizens who had not lived through the global wars of the twentieth century), the then Prince of Wales compared the Pick for Britain initiative with 'that great movement of the Second World War', the Land Army. 'It will be hard graft', he conceded, but 'hugely important, if we are to avoid the growing crops going to waste.'[5] In the UK, the National Farmers Union statistics show that in 2018 only 1 per cent of harvest workers were British nationals, with the vast majority coming from Eastern Europe.[6] The UK's exit from the European Union, which promised to restrict freedom of movement from December 2020, following the 2016 Brexit referendum, made these seasonal jobs even less accessible to the foreign workers on whom the harvest relied. The National Farming Union (NFU) warned that without them, one-third of good food could go to waste, left to rot, leading to increased food prices in supermarkets.[7] The Pick for Britain scheme, though, has largely been seen as a failure, as too few people signed up. Instead, the UK had to fly in Romanian workers in an emergency effort to rescue crops from rotting in the fields.

As these workers rushed to stop food supplies from going to waste, panic stockpiling led to waste in the kitchen. 'My husband bought lots of tinned foods' in case of supply chain issues at the start of the pandemic, one survey participant told me. 'His stock-piling stopped when a) it was clear there were only intermittent disruptions and b) when ou[r] cupboards were literally bursting with tinned peaches & rice pudding that I ended up throwing out at a later date.'[8] Likewise, at the time of writing my dad – the type of person to eat stale crisps rather than see them go to waste – is still somewhat grudgingly making his way through jars of 'dis-gusting' cheap instant coffee that my mum stockpiled at the start of the pandemic. Generally, however, as these shockwaves hit the nation's food supply, the value of food became more apparent and we wasted less of it. During the first four months of lockdown, from March to June 2020, the UK threw away at least a third less food than it had previously.[9] In our homes, food waste was reported at 24.1 per cent in November 2019, and this became 13.7 per cent in April 2020 during the first lockdown, meaning that 43 per cent of food that was once wasted was saved! In another national poll, 41 per cent of people reported wasting less food.[10] It seems, then, that the impacts of the pandemic on food waste throughout the food chain, from the fields to the dining table, were many and complex, and, all the while, Brexit was transforming the food supply system.

Leftovers in Lockdown

I care a lot more about not wasting food because suddenly I'm being limited to how much I can buy. Makes me realise how wastefully I was buying and eating before too.

Anon., 22 March 2020.[11]

I have become much more conscious of food waste as I am always at home.

Anon., June 2021.[12]

'From this evening I must give the British people a very simple instruction – you must stay at home.' We knew it would likely come, having spent weeks and months hearing about nothing else on the news except the relentless progress of Covid-19 across the globe, but the prime minister's words still felt heavy and illusory. There was a weighty sense that we were living through a historic moment. With clenched fists, Boris Johnson told the nation that we could leave the house only to carry out 'one form of exercise' per day (and only with members of our household), for a medical or care need, to travel to and from work if 'absolutely necessary', and – importantly for our purposes – to shop for 'basic necessities, as infrequently as possible'.[13] For many – including the secretary of state for transport, Grant Shapps – this translated as shopping for food once a week only.[14] Even before lockdown was announced, just going to the supermarket could be a slow and daunting task, since customers commonly had to wait in long queues standing two metres apart, and fewer people were allowed into each shop at a time. Abruptly, the surfeit of places from which we would normally buy food – restaurants, pubs, schools and offices – were closed down. As a nation, about 30 per cent of our calories were normally found outside of the home; suddenly all these calories had to be sourced for home consumption, normally from the supermarket.[15] Without those quick dashes to the corner shop to grab the missing eggs for a recipe, or a rushed sandwich from the mini supermarket on a busy workday lunchtime, we became much more conscious of the food that we had in our homes and how to use it effectively. As one anonymous contributor to my coronavirus food survey described in May 2020:

> I no longer pick up lunch/snacks when I'm out so I have a lot more knowledge of what I'm actually eating. I do still buy ready-made snacks but it's easier to only eat them some of the time when I can prepare my own food. I also used to eat out

quite a lot and that's obviously no longer happening so I guess I'm spending more time thinking about what I eat since I have to cook most of it.[16]

Almost four in five of us reported doing extra things to manage our food during the first lockdown, with, for instance 41 per cent thinking and planning more before going to the shops, such as checking the fridge and cupboards so as to purchase only those food items truly needed.[17] Indeed, another person told me from lockdown in April 2020 that they were 'inclined to use up the food we have in to minimise the number of trips to the supermarket', which would normally have been daily or at least every other day, but which had been reduced to every eight to ten days.[18] Where we shopped also changed. As certain items became less available in supermarkets, more people tried smaller local shops, like a local bakery or plastic-free shop for flour, the butcher for meat, or the milkman who could supply regular milk directly to the doorstep.[19] With heightened concerns about food waste and food packaging waste, others specifically sought out more sustainable delivery options, like Oddbox, a service that delivers fruit and vegetables from farms that would otherwise go to waste on account of surplus, size or look.[20] More than a quarter of people in one national survey reported shopping at small grocery shops more than they had done pre-pandemic.[21] Awareness of these businesses, which have also become more accessible as more of us work from home, and a desire to support them, seem to have created a greater sense of loyalty and may have changed shopping habits for good.

A drive towards localism and self-sufficiency continued at home. 'I'm currently in the process of buying seeds and cultivating the garden to grow my own vegetables, which I'm not sure I would have considered otherwise,' Lourens (then twenty-four) told me from lockdown in March 2020.[22] Like Lourens, many people reported taking up growing their own vegetables in a response to raised concerns about food supplies, a desire to improve their health and as a recreational activity given that it was no longer possible to socialise as normal. This was practised more commonly by people who were on benefits, and also among younger age groups (26 per cent of eighteen- to twenty-four-year-olds grew more of their own food compared to their pre-pandemic habits, as well as 20 per cent of those aged twenty-five to thirty-four), probably because they on average had more free time during the pandemic, were more likely to work from home, and had more money worries than older age groups. Conversely, those on the highest incomes also grew more of their own food.[23] These home gardens were not always successful... The dog ate Lillian in Southampton's harvest of gooseberries, and Jordon from Coventry found his broccoli had been ravaged by caterpillars.[24] Still, carrots, lettuce, strawberries, rhubarb and radish were among the fruit and veg proudly cultivated in lockdown gardens or on spare windowsills. Those who had not grown their own food before, but who were used to the abundance found on pre-pandemic supermarket shelves, were served just a taste of the labour that goes into putting food on our tables before we even cook it.

Back in the kitchen, the pandemic also altered the way we cooked. Suddenly we became much more conscious of the perishability of food, checking when food needed to be used up by, and often planning meals to make sure supplies lasted for as long as possible with less being wasted. As one contributor to my survey reported in March 2020: 'The food shortages have made me think more practically about what I eat in terms of prioritising perishable food above non-perishable.' This meant, for instance, using up fresh tomatoes rather than the tinned tomatoes in the cupboard, which could instead be stored in case they were needed for future use in the event of fresh produce shortages.[25] Having to come up with unusual ways of combining the ingredients in our fridges and cupboards without a spontaneous trip to the shops for specific ingredients, 40 per cent of us cooked more creatively during the lockdown.[26] Batch cooking became more common, and with this our freezers put in more work as we saved pre-cooked meals for later on. In the first three months of lockdown, the growth in sales of frozen foods, with their long shelf-life and value for money, was double that of the previous three months, booming by £285 million.[27] Meanwhile, in the first lockdown, 30 per cent of us started using up more leftovers than we had before.[28] As well as simply reheating meals, leftover cooking could become a creative endeavour. The leftover or stale ends of a loaf might be saved as croutons for soup or leftover sourdough starter might become the next day's pancakes.

The desire to 'eat every last scrap' was commonly reported regardless of financial status as many feared for their futures, but for those whose work was affected, thrift became even more important. Living off cheap own-brand tins, root vegetables and bulk grains like couscous, with a budget of £20 a week, Fabiola remembers wasting nothing, making sure she used even the leaves or roots of cauliflower and broccoli in a broth.[29] Financial insecurity was one reason why a significant minority turned to eating food past its use-by date in the first lockdown.[30] This drive to use up food sometimes also meant eating foods that were previously disliked, as was the case with Dorothy, who through 'a shift in mindset' as her income came under threat, taught herself to like sweet potatoes and frozen peas.[31] When scarcity and uncertainty threatened, 'waste not, want not' seemed once again an appropriate maxim.

Hunger and Help

If the lockdowns encouraged some to adopt a healthier diet, for many others the lockdowns had the devastating effect of pushing them into food insecurity. Since the 2008 financial crisis and the austerity policies of the Conservative government in the last decade, food bank usage was increasing exponentially even before the pandemic hit. The Trussell Trust, a charity which runs the majority of food banks in the UK, reported a staggering increase of 74 per cent in its provision of food parcels between 2015 and the start of 2020. 'Low income', 'benefit delays' and 'benefit changes' are the primary reasons given for referral to food banks.[32] Incredibly in the world's fifth-richest country, 2,000 food banks gave out food to 1.6 million people in 2018, and in 2019 those classified as living in poverty amounted to 22 per cent of the population.[33] Yet, as more people were thrown into financial insecurity and/or unemployment as a result of the lockdown, the number of food parcels distributed from food banks increased by 81 per cent between April and June 2020.[34] In June, 10 per cent of the population were using a food bank or food charity, and in July so were nearly one in four young people (sixteen- to twenty-four-year-olds) and 18 per cent of families with children.[35] In May 2020, the Independent Food Network, which represents food banks other than those run by the Trussell Trust, reported an almost 300 per cent increase in demand.[36] These statistics – which continue to shock me even now I am familiar with them – likely do not even represent the true scale of hunger, given that some preferred to skip meals rather than face the perceived stigma surrounding food bank use.[37] In late March 2020, 1.5 million adults in Britain said they could not obtain enough food,[38] and by September 4 million people, including 2.3 million children, had experienced moderate or severe food insecurity.[39*] Financial troubles were not only brought about by job losses or furlough, but since more people had to look after children or other dependents at home, and as a result of rising utility bills, which were exacerbated by staying at home rather than going into

---

* This equates to 14 per cent of households, an increase from the 11.5 per cent of households in this position before the pandemic.

work. Those who relied on a meal a week from a family member or friend – like a shared Sunday roast – suddenly had this dinner removed from their routine, and budget food shopping – seeking out reduced items, shopping around for deals, or going to specific budget shops – was no longer possible given the restrictions on our shopping habits.[40]

Abbie's story is just one from the millions who needed to rely on food banks during the pandemic. A single mum living in London, Abbie's two children were two and fifteen when the first lockdown was announced. Having had to leave her job after fleeing domestic violence, she was on Universal Credit and had just £60 a week to feed her family. Because of the pandemic, Abbie could no longer rely on her mum to provide meals at the weekend, and to avoid public transport she had to shop at a supermarket and corner shops that were more expensive than her usual stores. A £9 voucher a month from the Healthy Start scheme helped, as did the £15 weekly vouchers in place of her son's meals at school, which the Manchester United footballer Marcus Rashford successfully pressured the government to extend over the summer holidays of 2020 after they withdrew this often essential support. But Abbie still needed to supplement the family's food intake with supplies from food banks about once a month. She was regularly hungry during the lockdown: 'I'm the leftover queen so I make the kids their food, they eat and then whatever's left, I sort of cobble together and eat myself,' she said.[41]

As well as food banks, those who needed help getting food could turn to a variety of local food-sharing schemes, many of which relied on supplies from FareShare, the UK's largest charitable redistributor of food. Unlike food banks, which rely on donations from individuals, FareShare collects surplus food from farmers, packers, manufacturers and retailers. Products might have been rejected from supermarket shelves owing to a packaging error, failure to pass quality control (for example, 'oversized' vegetables), or simply be overproduced, either because they did not sell as well as expected or as an inbuilt surplus that shops order so as to not disappoint customers and that farmers and manufacturers produce so as not to disappoint supermarkets. While at the start of the pandemic, as people panic-bought supplies, donations from supermarkets to FareShare declined, when the hospitality sector closed down the charity was inundated with food stocks that were going to go off if not used up.[42] Supermarkets then pledged to donate more of their food supplies to FareShare and food banks via the Trussell Trust.[43] Local businesses also contributed surplus to redistribute to those in need. For example, the South West branch of FareShare received weekly donations from Samworth Brothers, food manufacturers who own the famous Cornish pasty-makers Ginsters, which in the pandemic year starting May 2020 equated to over 25,000 meals.[44] On a smaller scale, Naomi, who worked in a Birmingham office, described how her employer continued to order their weekly fruit

Figure 36. FareShare South West workers helping in the non-profit's massive pandemic drive to redistribute surplus food to prevent it from going to waste and to feed those experiencing hunger.

delivery from the greengrocer during the pandemic to support this local business, but donated it to a food bank to prevent it from going to waste.[45] In addition to its surplus food model, to help meet the growing demand, FareShare received special governmental funding, which lasted until March 2021. In the South West regional centre, the charity distributed an additional 642 tonnes of Defra-funded food.[46]

Phoebe Ruxton, director of fundraising and communications at FareShare South West, told me how, despite receiving no government funds in normal times, the charity's infrastructure was relied upon as the main avenue for distributing emergency food during the pandemic. Inundated with demand from new charities seeking food via their membership scheme, they very quickly launched an emergency warehouse, having to work on a case-by-case basis to ensure that each charity received the types of food they required before it went to waste.[47] In what was a monumental organisational feat, in total during the first wave of the virus between 16 March 2020 and 31 July 2020, FareShare South West distributed over 2 million meals to more than 300 organisations, which equalled more than six times the amount they had delivered pre-pandemic.[48] Over a year this equated to 6.1 million meals, which saved 2,985 tonnes of CO_2 emissions in wasted food, and supported 416 charities, schools and community groups.[49]

The 'Cheers Drive' project is one example of the many community initiatives that FareShare South West was able to support during the pandemic. A collaboration between chefs whose restaurants were closed as a result of the pandemic and Caring for Bristol, which works in the homeless sector, 'Cheers Drive'* aimed to redistribute food to Bristol's homeless people. Josh Eggleton, the owner of several local restaurants, explained that the enterprise started when he cooked up the surplus from his fish and chip shop, which was otherwise going to have to be

---

* Named after a well-loved Bristolian phrase used to thank the bus driver.

thrown away, and distributed it to the homeless people who were being temporarily housed in hotels by the council. The initiative quickly expanded, with Josh mobilising four kitchens across the city, collecting and using the leftover food from recently closed restaurants, their suppliers and then from FareShare. The team made 1,200 meals for 400 homeless people over a period of about a year. 'You wouldn't believe how much of it there was,' Josh tells me in reference to the surplus food they gathered, and the team 'didn't waste a damn thing'. All this took a great deal of coordination between restaurants, suppliers and other charitable projects. If Josh had collected a glut of 800 yoghurts, for example – too many to redistribute as part of the Cheers Drive project – he would get in touch with Tess and Elliot Lidstone at BOX-E restaurant in Wapping Wharf, who could use them in their food box scheme, in which they distributed food to sixteen- to twenty-five-year-old care workers.[50] Other organisations, like Muslims 4 Bristol and the Food Hub Consortium Project,* worked to make sure that any food that was donated and redistributed was culturally appropriate to those who received it, for example by ensuring that it adhered to halal laws.

In Bristol, new FOOD (Food On Our Doorsteps) Clubs were set up during the pandemic, at Broomhill, Oldbury Court and Speedwell, using food primarily gathered from Fareshare South West.† Those with young children under five could sign up to receive a full bag of food (worth around £20) each week at a cost of £3.50. This was especially important when schools were closed, as many members relied on free school meals to keep their children fed. The statistics of child food poverty in the UK are

---

* The Food Hub Consortium Project was established in April 2020, made up of Bristol Somali Resource Centre, Bristol Horn Youth concern, Talo, Bristol Black Carers, Bristol Somali Kitchen, Barton Hill Activity Club and Malcom X Community Centre.
† In partnership also with Early Years and Children's Centres services and Feeding Bristol.

staggering. In Bristol, 22 per cent (14,250) of our children expe-rienced hunger in 2020.[51] Gemma is a mother of four and started using FOOD Club during the first lockdown, when, without a car, she needed to collect food locally. 'My two older girls love getting the random assortment of food each week', she says, as they were able to look up creative recipes to make together, like a 'delicious curry' from butternut squash and chickpeas. 'It's bril-liant to hear that we are saving good food from going to waste,' Gemma expressed.[52]

These are just a few examples from my home city. Similar community projects blossomed across the country, as ordinary people came together to stifle the wastage of good food and use it to help prevent others from going hungry in a time of crisis. Even before the pandemic, however, food aid in the UK was moving from an 'emergency' provision to a model focused on 'routine' food insecurity.[53] Since 2010 successive governments have diminished both their budgets and the responsibility they are prepared to take for managing food poverty. While some argue that independent initiatives of the sort I have just described are a useful means of redistributing food surplus, many experts maintain that they should not take the place of state welfare. It can be extremely costly for charities and projects to find the resources, including vans, and fridges and freezers, to connect surplus food with those in need, all using a workforce of volunteers. And simply redistributing food rather than dealing with the underlying causes of food insecurity works to exacer-bate cycles of poverty. Surely we should expect policy makers in the UK – the fifth-richest economy in the world – to do more to ensure that its inhabitants have access to sufficient amounts of food? Many survey participants likewise commented that in an ideal world there would be no hungry bellies for the community to feed, while the food system would create very little surplus at all. As Jordon told me: 'We must avoid the situation of leftover food for leftover people.' [54]

A Leaky Food Chain

In August 2021, McDonald's ran out of milkshakes and Nando's ran out of chicken, forcing the temporary closure of fifty restaurants. Labour shortages caused by rising Covid cases impacted on the food industry across the globe. As one article from September 2021 put it, 'the global food ecosystem is buckling due to a shortage of staff'.[55] This came in the midst of the so-called 'pingdemic' in the UK, when high numbers of people across the food industry were told to self-isolate by the NHS Covid-19 app after having come into contact with someone with the virus, leading once again to bare supermarket shelves. And shortages were further exacerbated in the UK by Brexit, which officially took place on 31 January 2020, as the flow of EU workers dried up.* By March 2021 there were 16,000 fewer EU nationals working as HGV drivers than in the previous year. A Road Haulage Associated survey estimated a shortage of 100,000 drivers, out of a pre-pandemic total of 600,000.[56] This led not only to a paucity of certain foodstuffs, but an increase in cases of food wastage. In September 2021, for example, there were reports that a number of dairy farmers were dumping milk. Sky News spoke in early October to a dairy farmer in central England who said they'd thrown away 40,000 litres of milk which hadn't been collected since August as a result of HGV driver shortages.[57] Andrew Mellot, a dairy farmer in Dilhorne, Staffordshire, also got a call on 10 September to say that the 1,600 litres of milk he had produced that day would have to be disposed of because it couldn't be collected. He said: 'It is a complete waste of a good food source, especially in these uncertain times when we hear there could be shortages.'[58] Retailers often stipulate in contracts with their suppliers that the milk can't be sold to anyone else,

---

* HGV shortages were also felt in parts of Europe, but they were able to call on drivers from across the EU Single Market.

meaning that many dairy farmers were left with little choice but to throw it away.

In early October 2021, reports emerged that over 100,000 pigs on British farms would have to be culled as a result of the huge shortfall in abattoir workers. The British Meat Processors Association records that around 80 per cent of staff in two major processing centres in Hull came from Eastern Europe, but in the wake of Brexit, many of these workers left the UK. This pattern of movement intensified as a result of the pandemic. Meryl Ward, the owner of a family farm in Lincolnshire, said in October 2021 that she already had 1,600 pigs that should have gone to slaughter but were instead using up expensive food and resources on the farm. Without the necessary processing facilities, the options for productive use of the livestock were vastly limited: some of the flesh might be sold on for cheap lard or pet food, but most of the animals would simply be incinerated. Meryl described how 'producers are in despair', adding: 'We can't just waste this food. It's criminal.'[59] The government's emergency introduction of 800 temporary foreign visas was too slow to halt the cull, only 100 foreign butchers taking up the offer. By December 2021 more than 30,000 pigs had been needlessly killed.[60]

Especially in the east of England, many farmers voted to leave the European Union because they believed that their way of life was being eroded by rural migrant workers.[61] Yet, without workers from the EU, who made up one-quarter of all UK food industry workers in 2018,[62] the economy still needs to find migrant workers to fill a number of food processing roles from fruit pickers and abattoir workers, to cooks and waiters, to ensure that our food moves from farm to fork. A letter signed by the National Farmers Union and the Ulster Farmers Union, several supermarkets, and significant food businesses like Dairy Crest and Muller, urged the government to keep Britain in the European Single Market in order to maintain the flow of migrant workers and tariff-free access. However, their pleas fell on deaf ears. In 2020, the UK exited the EU in a so-called 'hard Brexit', which entailed leaving both the Single Market and the Customs Union.[63]

Meanwhile, the combined effects of Brexit and the pandemic were leading to food waste in the fishing industry. The vast majority – over 90 per cent[64] – of people in fishing communities voted for Brexit, many believing that this would counter the restrictions of the EU's controversial Common Fisheries Policy (CFP), which since the 1970s has given each member state equal access to EU waters. British fishermen wished to gain a greater share of lucrative British waters, which make up one of the largest Exclusive Economic Zones (EEZ) in Europe. Within this system the EU set quotas of the amount of particular fish that trawlers could bring to land in an attempt to limit overfishing. This policy was particularly contentious, however, because it led to a huge amount of waste as fishermen threw overboard fish that they had accidentally caught but which took them over the quota. They might even cast away – dead or alive – smaller fish in favour of more valuable bigger specimens. Iceland and Norway in part refused to become members of the EU as a reaction to the Common Fisheries Policy.[65] In 2020 Conservative MP and former Environment Secretary Owen Paterson claimed that fishermen in the EU were 'throwing back a million tonnes of fish dead as pollution each year' as a result.[66] Brexit would seemingly lead to the end of this wasteful policy. Yet, starting in 2015, the EU had introduced a 'landing obligation', which made fishermen keep certain species of fish that they caught, and by 2019 this meant all fish to which a quota applied. Like the EU, after Brexit the UK government continues to ban the discarding of quota species (with an update to this effect in 2023).[67] Instead of the imagined reduction in wasted fish, however, in the first few months of 2021 reports began to appear of British-caught fish stocks rotting before they were able to cross the Channel to enter the European market. The problem lay with increased bureaucratic checks between the UK and the EU post-Brexit, leading to enormous transport delays. EU buyers began to reject British catches. These losses, aside from the waste of animal lives, struck devastating financial blows to those economically reliant on the fishing industry, which in 2019 had sent nearly three-quarters of its fish exports to the EU market.[68]

As well as restricting exports, Britain's departure from the EU in 2020 meant ultimately turning away from our nearest food source, at a time when the EU supplied 30 per cent of the food eaten in the country.[69] An additional 11 per cent of the UK's food imports from further afield were processed in Rotterdam in the Netherlands. In total, Britain imports double the amount of food that it exports. Although there was very little discussion about the likely impacts of leaving the EU on the food supply in the run-up to the referendum, Brexit rocked – and continues to redefine – the entire UK food system. Importantly, Brexit exposed the problem with the 'just in time' system in which very little is stored in ware-houses and is instead delivered just before it goes to the shop shelf. In this way, there are very low stocks of food in the country at any one time, exposing Britain to waste and shortages in the increas-ingly likely case of delays at the border as more checks are required. The EU introduced full border controls in January 2021. Britain, however – beyond requiring the completion of a customs declara-tion on food coming into the UK from the EU – postponed the introduction of further food controls several times after March 2021. When he scrapped plans for additional checks on EU food imports in April 2022, the minister for Brexit opportunities, Jacob Rees-Mogg, called them an 'act of self-harm'.[70]

Brexit Bites

So as the country's food system struggled to deal with the coronavirus pandemic, what vision did those in charge have for Brexit Britain? And how do these plans impact on food supply, waste and surplus? Many touted 'hard Brexit' as an opportunity, instead of trading with the EU, to make new food deals with countries further afield. Famously (or infamously, depending on your perspective), in June 2020 Boris Johnson plugged the UK's first post-Brexit trade deal, which was finally signed with Australia in December 2021, as an opportunity to receive tariff-reduced Australian chocolate bars: 'How long can the British people be deprived of the opportunity to have Arnott's Tim Tams at a reasonable price?' he joked.[71] Likewise, the post-Brexit trade deal with New Zealand, signed in February 2022, will allow Kiwi farmers to export lamb and other comestible products all the way to the UK with reduced and removed tariffs, which will mean that they can be sold to UK customers at a lower price. From the late nineteenth century, Britain pursued a policy of free trade, seeking out cheap produce from its colonies and trading partners across the globe, rather than focusing on domestic food production. The new 'global Britain' imagined by the UK government appears to hark back in its rhetoric to this imperial era, a seemingly more lucrative time in the country's history before more protectionist measures were applied.

Though of course it would not be feasible to avoid food imports altogether, there are issues with a dependence on the globalised supply chains that the UK government hopes to (re)build in the light of the Brexit vote. As Tim Lang, Professor of Food Policy at City University, wrote in 2020 during the pandemic, the UK no longer employs the military or naval power to protect food stocks as they are brought in from abroad. In 2017, in the immediate aftermath of the referendum, it emerged that the UK Border Force had just three active vessels to protect its borders. Although this number was increased to five, it compares unfavourably with Italy's fleet of 600 or the 240 such ships that Greece possesses. 'Cabinet ministers (of many political persuasions) still naïvely believe', Lang wrote, that the UK can rely on food grown on other continents 'as though [...] the UK still "rules the waves"!'

Depending on faraway countries to grow our food, especially as climate change threatens agriculture in many of these areas, ultimately risks us losing access to food supplies.[72]

Lang argued, instead, that Brexit offers an opportunity to rethink our food system, and, importantly, to grow more of our own food. Indeed, the Brexit vote, along with Trump's election as US president in the same year, 2016, were often explained at the time as arising from a sense among the population of 'being left behind' by globalisation. Despite many British farmers having voted to leave the EU, they have expressed frustration at the Australian and New Zealand post-Brexit trade deals, which may result in them being unable to compete with cheap foreign imports. Moreover, according to the agricultural scientist Charlie Clutterbuck, unchecked

Figure 37. Bagging export lambs at Southland Frozen Meat Co. Ltd, New Zealand, c. 1960. New Zealand's surplus sheep stocks were frozen and exported to Britain from the late nineteenth century; Southland Frozen Meat Company was established for that purpose in 1881. The Brexit trade deal means more of New Zealand's lamb will be bought by UK consumers once again.

free trade leads to waste as all farmers experience a good harvest at the same time. The market is flooded with a certain product, and it must be sold for a cheaper price as a result – if not simply discarded – while farmers invest in cutting costs further to keep afloat. Instead, Clutterbuck argued in his book *Bittersweet Brexit*, written shortly after the referendum, that British farmers should diversify, moving away from monocultures so that there is less opportunity for gluts of particular crops to accumulate, while the market as a whole closes off to outside competition. We could grow 2,000 varieties of apple in the UK, he noted, but currently we see only a handful of different types on supermarket shelves.[73]

To protect British-grown food the UK industry has relied on government subsidies since the 1930s. The EU's Common Agricultural Policy, which gave out subsidies to all EU farmers, was one of the most resented features of EU membership. In the 1970s and 80s, as we've seen, the policy led to wasteful food sur-pluses, dubbed 'milk lakes' and 'butter mountains' in the press, though reforms that placed caps on output have since overturned this situation. Forty per cent of the EU budget goes to land-owners in the subsidy scheme, in a sense to prevent them from producing more food than the market can handle while allow-ing them to maintain a living.[74] Outside the EU, the UK has an opportunity to redistribute the CAP subsidies, which amounted to £3 billion. Clutterbuck suggested they might be used to fund farm labourers rather than landowners as a way of cutting the costs of food production to ensure that food prices are kept low while improving working conditions in the agricultural sector.[75] The failure to recruit British workers to rescue crops from being wasted during the coronavirus pandemic certainly demonstrated that agricultural labour conditions will need to improve if a more local, self-sufficient food system is desired post-Brexit. While it has temporarily extended the current subsidies in a transi-tion period of seven years from 2021, in the Agricultural Act of 2020 the UK government set out the legal framework in order to distribute funds to landowners based on their contribution to 'public goods', for example, for improving environmental or animal welfare standards.[76]

———•———

If we as a society had taken food for granted before, the pandemic made its value acutely obvious. The global infrastructures that keep our supermarket shelves full started to buckle under the pressure of staff shortages, transport delays and increased demand. The so-called 'just in time' system that the supermarkets rely on in Britain means that very little food is stored in warehouses. Instead, an international network of logistics keeps 50 per cent of our food supply coming in from abroad, and at least 30 per cent of it arriving from the EU.[77] At the same time, the UK's uneasy exit from the European Union, its source of workers who picked, harvested, slaughtered and transported our food, and its Single Market in which to easily trade foodstuffs, threatened food supply chains further, leading to food waste in farms, in transit and in warehouses. Questions about where and how we would secure our future food supplies were still far from answered as the UK left the EU in January 2020, four years after the referendum vote, on 'hard Brexit' globalist terms. In the confusion of shortages and fear of multiple coronavirus lockdowns that forced us to stay within our homes, we thought about food – buying, making and consuming it – more often. For some, it could be a comfort, something that provided structure in days that otherwise blended into each other, while for others, those with financial problems or eating disorders, for example, it could be a source of seemingly non-stop anxiety. In a national survey 67 per cent of respondents said that they felt differently about food as a result of the coronavirus pandemic, with 38 per cent maintaining that it was the most significant event in their lifetime in terms of their relationship to food and food waste.[78]

The food system exposed by the coronavirus pandemic was an unequal one. Yet to help those unable to access food, the pandemic inspired a strong community response. For example, four out of ten people helped others by shopping for food for those neighbours, friends or family members who were isolating.[79] This was particularly common during the first lockdown, when online supermarket delivery slots were difficult to secure given the surge in demand placed on systems that had not been built to handle these levels of traffic. As we've seen, initiatives that collected and redistributed surplus food or food that would otherwise have gone to waste to those experiencing food insecurity sprang up rapidly in local communities with the support of major food redistribution charities like FareShare and the Trussell Trust. The government also responded by distributing emergency food parcels to the most clinically vulnerable people who had been asked to isolate or 'shield' in their homes over the first lockdown, giving the contract (controversially without tender) to the wholesalers Bidfood and Brakes, who had superfluous stock following the shutdown of the hospitality industry which they usually supplied.

When the coronavirus lockdown first came into effect, the Food Ethics Council, a think-tank and charity, urged us to use this opportunity to reimagine a fairer food system, in terms of its social, economic and environmental impacts.[80] Bristol's successful bid to win a Gold Sustainable Food City award (the second to be awarded to a UK city after Brighton and Hove), offers an important example of how the pandemic could inspire top-down reform. The councillor Asher Craig, who chaired the Going for Gold Steering group, declared that 'the onset of a pandemic did not deter our ambitions, instead it has been a catalyst for change'.[81] The key themes of the change sought included: 'buying better', by relying on locally produced food; 'eating better', meaning nourishing good-quality food businesses; 'food equality', so that food is accessible to all; and 'zero food waste'. The 'Bite Back Better' campaign was launched in November 2020 during the second Covid lockdown, with online webinars on various sustainable eating habits, virtual community feasts and a website with food-saving tips and activities. Visitors to the website can print out a food planner to share with flatmates to manage items in the fridge, or learn about how to cook up banana skins (which are edible and full of nutrients!).[82] Still, the targets set by the group are a long way off. By 2050, Bristol aims to have completely eradicated avoidable food waste, and to be recycling all non-avoidable food waste.[83] Other local councils and the national government will need to adopt similar schemes if the UK's food system is to be significantly reformed. The UK's Environment Act, which laid out post-Brexit environmental goals, was finally passed into law in November 2021 after over three years of delay. It states that every household must have access to weekly food waste collection, a change that was due to be enforced 'by 2023',[84] but deadlines are as yet undefined for local councils with existing waste disposal contracts given that it takes time to amend them.[85]*

The true impacts of the pandemic and Brexit, which accompanied it in 2020, are yet to be fully understood or felt. Should we write off the food waste issues of Brexit as necessary teething problems, as the then prime minister Boris Johnson suggested?† Many trade negotiations remain in an 'adjustment period'. The fishing agreement of late 2020, for instance, arranged for 25 per cent of the existing EU fish quota in UK waters to be transferred to the UK over the period until 30 June 2026, after which time the

---

* A caveat that more than one type of waste can be collected together when 'it is not technically or economically practicable' also means that in practice household food waste may still be collected alongside garden waste in some areas. UK Government, Environment Act 2021, 57 'Managing Waste', 45A(6).
† Boris Johnson speaking about the pig cull on BBC One's *The Andrew Marr* show, 3 October 2021: 'I hate to break it to you, Andrew, but I'm afraid our food processing industry does involve killing a lot of animals, that is the reality [...] There will be a period of adjustment, but that is, I think, what we need to see in this country.'

number of Total Allowable Catches will be negotiated each year.[86] Trade deals to secure food supplies can take years to come to fruition, and the UK's farming industry is still in a state of flux. As of 2023, with war in Ukraine affecting the global supply chain, food shortages and limits on certain items have been reported once again.[87] The cost of living crisis continues to push more Britons into food insecurity, forced to rely on food banks and donated food surplus to survive.

If these wider economic and political movements seem out of our hands, will we see long-term changes in our everyday food habits at home? Even within that first year of the lockdowns, our initial thriftiness relaxed. While reported food waste in UK households was at a low of 13.7 per cent in the first lockdown, it rose to 18.7 per cent in November 2020 when England was in its second lockdown. Though still lower than the 24.1 per cent pre-pandemic figure of a year previously, by summer 2021 self-reported food waste was back in line with 2018 levels.[88] It is easy to see how less time spent at home can lead to increased household waste as people have fewer opportunities to manage the contents of their fridges and cupboards. A return to pre-pandemic 'normal' life, too, means for many tiresome hours stuck in commuter traffic jams and rushed school pick-ups that leave little time to think creatively about using up last night's leftovers. So what will the future of food waste look like, post-pandemic and post-Brexit? In the final part of this book, I'd like to pull the past into the present, to explore the initiatives being developed that might help us to create a food waste-free future.

EPILOGUE

A FOOD WASTE–FREE FUTURE?

The coronavirus pandemic appeared momentarily to have stifled concerns about the escalating environmental catastrophe and the role of the food system within it. When coffee shops began to open again, I remember having to unlearn my hard-earned habit of bringing my own reusable coffee cup with me, and instead to use a disposable version; the former were no longer accepted given fears of the Covid-19 virus spreading on surfaces. As we look towards the future, however, there is no doubt that human-made climate change is the most pressing threat to our planet. We are at risk of losing thousands of unique and perfectly adapted species as their ecosystems collapse, as well as facing the prospect of widespread human migration and hunger as famine, soil erosion and other natural disasters displace populations and threaten global food security. Desertification, changes in rainfall patterns, new and increasing pests and diseases will all contribute to crop failures or declines, especially in the tropics and subtropics.[1] In the UK, 53 per cent of us recognise the link between plastic packaging and climate change, but far fewer – 32 per cent – are aware of the environmental impacts of food waste.[2]

Crops that go on to be wasted still guzzle up water in irrigation, fertilisers and herbicides (especially high-yield selectively bred examples like those we discussed in Chapter 6), and diesel in the form of fuel for farm machinery. Discarded meat depends in turn on the grain used to feed livestock, which, as well as the energy needed to produce their food, use up land and pharmaceutical drugs in their lifetimes, all while emitting methane into the atmosphere. The land used to pasture animals and grow crops is robbed from natural landscapes such as carbon-absorbing forests. Energy is also needlessly expended in processing, packing and delivering food to customers, in the UK much of it from far-away countries. The estimated one-third of all food produced that ends up as rubbish is responsible for 8–10 per cent of greenhouse gas emissions, and as it rots in landfill it further pollutes nearby waterways and groundwaters.[3] Wasting less food and food packaging, and making better use of that which is disposed of, will not only save us money during insecure times, but will necessarily play an important part in our attempts to soften the blow of impending environmental disaster.

At the same time, our planet's resources are being put under further strain as the global population grows. By 2050 the UN predicts that we will need 60 per cent more food to feed the likely more than nine billion people who will live on the Earth. This would equate to the appropriation of 170 million more acres of farmland and the intense exploitation of natural resources like water in order to produce more crops. Already, more than 800 million people globally, around one in nine of the current world population, are hungry or undernourished.[4] For many experts, however, cutting food waste rather than producing more food is the obvious solution to our growing collective waistline as the world's population grows. Cutting food waste by half could account for around 22 per cent of our future food needs.[5] As one of the targets of the United Nations Sustainable Development Goals, nations around the world, including the UK, have indeed committed to halving food waste and loss by 2030.

The systems that we create to grow, trade and eat food – and how wasteful they are – affect the most significant issues that face us today including climate change, but also ecological and animal welfare, human health and socio-economic inequality. So how might the UN achieve its goal? And how might the food waste that we do produce be more sustainably managed?

Progress, of course, relies on top-down initiatives. Food waste is a systemic problem and needs strong leadership to tackle it. In France and Italy, governments have already established laws to force supermarkets to donate unsold food rather than destroying it.* In the UK, the efforts of Tristram Stuart and other food waste activists did much to compel supermarkets to act on the amount of food waste they create. Since around 2014 supermarkets have been selling vegetables that don't meet their aesthetic standards as 'wonky veg' or redirecting it into their own-brand products, which means that they take more of their suppliers' stocks and produce less waste. In the same year, the passing of the Groceries Coach Adjudicators Act installed an independent regulator to ensure supermarkets are adhering to the 2009 Groceries Supply Code of Practice. This legislation made it illegal for supermarkets to cancel an order to a supplier, whereas before they had often worked by overestimating demand before cancelling at the last minute when their actual requirements became visible. Now Tesco has become one of the most progressive supermarkets in the world in terms of its work in food waste reduction, according to Stuart, who says it leads in asking its suppliers to report on waste.[6] Ninety-five per cent of the food retail sector has now signed up to the Courtauld Commitment[7] to help reduce their food and packaging waste, and signatories across the food industry are responsible for 93 per cent of all food sales in the UK.[8] After the public consultation first promised in 2018 was postponed because of Covid-19, there are renewed calls for all stages

---

* The French law came into effect on 11 February 2016, and Italy followed around six months later. Instead of enforcing fines as in the French system, in Italy the law offered tax reductions in exchange for donations of surplus food.

of the supply chain – farmers, manufacturers, distributors – to be required to disclose their food waste statistics so that these can be properly managed and so that customers can make more informed decisions about where they shop for their food. To really tackle the problem with the urgency it requires, shouldn't these targets be legislated rather than voluntary?[9] In summer 2023 the government scrapped legislation to enforce mandatory reporting for large and medium-sized businesses over fears that the cost to retailers might result in increased food prices, even after the government's own research found that a 0.25 per cent reduction in food waste would balance any costs, and after the much-delayed public consultation found that 80 per cent of respondents supported the law.*[10]

Local and national government action should also help consumers to better manage food waste in the household. Bristol was the first Core City† to introduce separate household food waste recycling collections in 2006.‡ This is still far from a universal provision across the UK, however, and councils need to be properly funded to provide such services after austerity measures cancelled their roll-out in many areas. Despite taking the time to be 'meticulous about separating our waste' during the coronavirus pandemic, Anne from London told me that she was worried 'that my borough doesn't actually do anything useful

---

* As this book goes to press, in November 2023, the government has promised to review this decision in the first half of 2024 following the launch of legal proceedings against it filed by Feedback. With the support of other leaders in the industry, the campaign group argued that the government's decision ignored expert advice such as that from the Committee on Climate Change, which had recommended that mandatory food waste reporting be introduced by 2022.

† This refers to Core Cities UK, an alliance of eleven major cities in the country: Belfast, Birmingham, Bristol, Cardiff, Glasgow, Leeds, Liverpool, Manchester, Newcastle, Nottingham and Sheffield.

‡ This was for kerbside properties only. The full roll-out for flatted properties was completed only in 2021.

with it and it just ends up in landfill'.[11] The government pledged to reduce the amount of food waste in landfill by ensuring every household receives a separate weekly food waste collection 'by' or 'from' 2023 (though in practice it's unclear when after this date it will be enforced), and councils must ensure it is then recycled or composted.[12] In Bristol, the council ran a 'Slim My Waste – Feed My Face' campaign, to encourage residents to reduce the amount of food waste they produce and to throw any away separately, but still, in 2019, only 45 per cent of household food waste was being put into the separate food waste bin.[13] WRAP's Love Food Hate Waste initiative has run several successful campaigns on the impacts of household food waste and how to reduce it, like the yearly Food Waste Action Week, which have reached a broad audience across the UK.[14] Yet, there is still much more that our leaders could do to reduce household food waste. Elsewhere, governments have adopted more radical policies, like in South Korea, which from 2005 banned food waste from going to landfill, and where households are charged based on the amount of food waste that they produce.* Consequently, food waste decreased by 30 per cent by 2015, and by 2019 95 per cent of the food waste that was created in the country was being recycled.[15]

As well as from households, the 2021 Environment Act promised to enforce separate food waste collections in industrial or commercial premises. From late 2023 in Wales, and at an as yet undecided date for England, businesses will only be able to send their food waste to composting or anaerobic digestion rather than landfill.[16] In England, this will end the use of macerators to mush up leftovers and dispose of the resultant dross in the sewers, a practice already banned in Wales, Scotland and Northern Ireland.

---

* In the summer of 2007, UK Prime Minister Gordon Brown attempted to institute taxes on discarded food but after facing opposition from local councils, the policy was dropped.

Lawmakers could also direct funds to support producers to make good disposal choices. Currently, for instance, FareShare receives no financial help at all from the government, despite the successful trial of its Surplus with Purpose scheme, which helped companies with the costs of redistributing their surplus raw food rather than sending it to animal feed, anaerobic digestion or landfill. For the proportionally modest sum of £5 million per year, FareShare argue, they could rescue 53 million more meals annually.[17] As the UK untangles itself from EU regulation, there is an opportunity – should the government wish to take it – to reimagine the entire agricultural industry in a way that reduces waste. Sustainable farming practices that move away from the monocultures and the heavy artificial pesticide and fertiliser use that began in earnest in the 1960s, will ultimately enable better soil quality to help prevent future food loss and ensure future food supplies. Modern farming practices are, ironically, likely to lead to declining land productivity. In animal rearing, Brexit offered some food waste experts hope that leftovers could be reintroduced into pigswill once again, after the 2001 foot-and-mouth outbreak led EU leaders to ban this practice, despite the fact that pathogens can be destroyed if it is properly treated. Ironically, the EU has since removed the restrictions on feeding pigs from processed poultry remains, while the UK is yet to act.

Beyond governmental initiatives, as the world searches for new ways of meeting the UN's goal of halving food waste by 2030, new technologies that reduce, reuse or redistribute food waste have become a buzzing area of investment. To reach the target, ReFED, a US-based non-profit working on food waste, estimates that $14 billion per year will needed to be spent in the US alone, while this will generate $73 billion a year in return.[18] Innovation has an important part to play in our quest to forge a food waste-free future, transforming the food system across production, consumption and disposal. The following sections will explore some of these promising enterprises.

'A Rotten Apple Spoils the Barrel'

When I was growing up, each autumn the tree at the bottom of my parents' garden would shed bucket-loads of apples. Sometimes they were scratched or bruised, a little misshapen, pocked with the occasional insect burrow and matted with a natural powdery wax coat. Nothing then, like the shiny round unblemished examples that we bought in the supermarket the rest of the year. Did you know that the natural coating is washed off when apples destined for the shops are cleaned to remove dirt and pesticides? The fruit is then often coated with an artificial wax, made from beeswax, sugar cane, shellac (made from the Indian lac bug), candelilla or carnauba (the latter two both being derived from plants*). This technique was first patented in the US by Ernest Brogden, who founded the Brogdex Company of California, in 1922. Customers now expect the shiny perfect appearance that artificial wax imparts, and so the process is carried out for aesthetic reasons,[19] but it is also a food preservation technique, which works by reducing the moisture loss and acting as a barrier to burrowing insects. The wax also often contains fungicides to prevent the growth of mould. It is approved as safe for human consumption, but still regularly becomes the subject of alarmist campaigns that warn consumers of this hidden enemy; perhaps you've seen one of the many viral videos on YouTube or TikTok in which an influencer shocks the audience by removing layers of wax from apples, some even claiming that it can cause cancer.† Those who worry about the artificial coating will

---

* Respectively a type of euphorbia and the carnaúba palm tree.
† This claim is likely rooted in the fact that the EU banned morpholine – an emulsifier used to enable wax to mix with water – because of a risk that it will oxidise into carcinogenic nitroso compounds when nitrates are present in the body. It is approved as a food additive in the USA, Canada, Australia, South America and elsewhere, however. See Diana I. S. Kolberg et al., 'Morpholine, Diethanolamine and Triethanolamine Prohibited Additives in the EU', (2012), https://www.eurl-pesticides.eu/library/docs/srm/EPRW_2012_PM_030_Morpholin.pdf.

be pleased to hear of new technologies that are helping to extend shelf life and therefore reduce the amount of food wasted to rot and mould. One of these inventions is an edible, tasteless coating that is sprayed or dipped on fruit and vegetables to at least double its shelf life. Dr James Rogers, a materials scientist in California, has been developing Edipeel, with his company Apeel Sciences, since 2012. The coating comes as a powder, which is then diluted with water, made from the molecules in discarded organic produce, including leftover grape skins used in wine fermentation, apple stems and seeds and grass clippings. This forms a seal that keeps oxygen and insects out, while trapping moisture in. Before the idea was born in his mind, James remembers reading an article which stated 'All fresh produce is seasonable as well as perishable'. This simple truth made him ponder the long-standing conundrum in which produce is 'either in season and [you] have more than you know what to do with, or you have nothing'.* With Apeel technology fruits maintained their quality and mass better than if treated with artificial wax.[20]

Another way of making produce last longer involves controlling the environment in which it is shipped, perhaps most obviously by keeping storage containers cold. In a process that began with the first refrigerated ships of the 1870s, today, the 'cold chain' brings us fruit and veg from across the world in refrigerated shipping containers on ships, planes and trucks. Often deployed in combination with refrigeration, Controlled Atmosphere (CA) storage works by manipulating the gases in the atmosphere that fruit and veg are stored in. Based on principles discovered in the 1920s, CA storage began to be taken up on an industrial scale for certain crops in the 1950s. Essentially, it works by increasing carbon dioxide and lowering oxygen levels to slow down the respiration rates of the fresh

---

* Cambridge Crops, based in Massachusetts, USA, also make an edible coating, which is made from silk proteins rather than the plant lipids that Apeel use.

produce, while stunting the growth of microorganisms. Modified Atmospheric Packaging (MAP) works to control gas levels within individual packages of food. One MAP solution comes from the US-based company Hazel Technologies, who sell packets that can be put into boxes of produce and alter the atmosphere to slow the food's response to ethylene. When an apple, or any fruit for that matter, starts rotting, it emits ethylene into the air. This in turn stimulates the ripening of fruit nearby: a rotten apple spoils the barrel. SmartFresh, owned by the US company AgroFresh Solutions, has since 2002 been selling a product with the active ingredient 1-Methylcyclopropene.* This is a newly created chemical, invented in the 1990s, which works by mimicking ethylene and blocking its effects at a cellular level within the fruit. Treated with this chemical, fruit can last up to a year before it is sold to consumers, when it is blasted with ethylene to reawaken it, though there is some degradation in its taste and appearance.

Rather than focusing on slowing down the ripening process, the innovative technology from Ryp Labs, a company that develops waste-saving food technologies, limits damage caused by pathogens like fungus and bacteria. The company has developed a sticker (as well as a sachet) containing a volatile compound which repels pathogens. Moody Soliman, the CEO and co-founder, tells me that this mimics the 'secondary metabolites' in volatile compounds emitted into the air naturally by plants to protect them from dangers. Fungicides and biocides are otherwise not applied to fruits and vegetables other than during the pre-harvest period, and the technology is particularly useful for non-climacteric fruits – like strawberries – which don't continue to ripen after they are harvested, since these cannot be treated with technologies like AgroFresh and Apeel. Using Ryp Labs stickers, strawberries lasted an extra two days before they started to degrade, which means that the supermarkets can accept a

---

* It was approved in the US in 2002, in the UK in 2003 and the EU in 2006.

much greater proportion of fruits from suppliers rather than them going to waste at this stage of the food system.[21]

AI Food Savers

The solutions described so far increase the shelf life of food, but a longer shelf life doesn't always prevent waste if consumers, supermarkets and retailers overestimate demand – and they routinely do. The modern food system is premised on the fact that our eyes are bigger than our stomachs, since, as we've seen, it is often most cost effective for retailers to order more food than is needed in order to avoid customer disappointment. To help tackle this problem at the supermarket level, companies such as Edinburgh-based Neurolabs are developing artificial intelligence algorithms that predict demand. They use information like weather, competitors' prices and the local economic makeup of an area to predict which products the supermarket should order in, and use video cameras installed in the shop to notify staff when the shelves need to be restocked.[22] AI is also being used to aid the common method of marking down food when it is coming to the end of its shelf life. Israeli startup Wasteless works on the premise that food closer to its use-by date should be reduced to encourage consumers to buy it and to avoid needless waste. It uses AI to automatically lower the prices of foods, which are then displayed on a digital label on the shelf. This, the team claim, will reduce the amount of food waste by 50 per cent. Supermarket revenues would also increase by 20 per cent as they could sell food (even at a discount) which would otherwise have to be thrown away.[23]

Restaurants, too, are benefiting from new AI systems. Winnow Vision uses a smart scale to log what kitchens throw away. A camera in the kitchen bin takes a photograph of what is wasted and the AI system has been trained to recognise the type of food that has been discarded. Knowing the weight and unit cost of the item selected,

Winnow's technology then draws up a daily report which helps chefs understand what types of foods they are wasting and therefore how best to run more efficiently and save money. In the hospitality sector most waste is made before it gets to the customer, from off-cuts, cooking errors and most significantly from overproduction. By highlighting areas of overproduction and how to fix them, the system expects to cut a kitchen's food purchasing costs between 2 and 8 per cent. When IKEA piloted the technology from Winnow Solutions in 2015, they found they were able to achieve around a 40 per cent decrease in food waste within just six months. Now, Winnow Vision is found in over forty IKEA stores in more than thirty countries. In IKEA UK & Ireland, this system has cut food waste by 50 per cent and saved 1.2 million meals.[24] Currently deployed in the larger kitchens of contract catering, retail, hotels, casinos and cruise ships, future improvements in Winnow's AI will mean that the streamlining technology will be available to smaller kitchens too.[25]

Figure 38. A commercial kitchen using Winnow Visions' AI technology, which measures the amount and type of food waste produced so as to better manage it.

Fridges of the Future

In the process of moving home in 2014, Tessa Clarke, the daughter of Yorkshire farmers, was troubled that the six sweet potatoes, whole white cabbage and pots of yoghurt she had left in her shortly-to-be disconnected fridge would have to be thrown in the bin. With co-founder Saasha Celestial-One, Tessa ended up creating Olio, an app where consumers can share their leftovers with each other. Now with over four million users, the app has saved the equivalent of 100 million car miles in emissions.[26] Serving its first meal in 2016 from its headquarters in Copenhagen, Too Good to Go is another app that allows customers to collect leftovers, this time specifically from businesses who have surplus food that would otherwise go to waste. After three-and-a-half years trading, the app saved 29 million meals from the bin, which is the equivalent of 66,000 tonnes of greenhouse gas emissions.[27] Other food-saving apps followed hot on its heels. Kitche, for instance, allows consumers to scan in their receipts to keep track of the food they have in the house, sending reminders if fruit, veg or other perishable items are at risk of going off. Users have access to thousands of recipes which can be filtered based on what foods they have in their fridge, while a 'tossed' section tracks what food has been wasted so that the user is conscious of their food waste habits. Similar examples are found across the world, like France's Magic Fridge app or India's Seva Kitchen.

A more high-tech solution to food waste in the home comes from London-based Mimica, who have developed caps and labels for use on food and drink packaging that change in texture (from smooth to bumpy) to indicate that the product has expired, using a temperature-sensitive gel. Inventor Solveiga Pakštaitė came up with the idea for her final year dissertation project at university design school while interning with The Guide Dogs for the Blind Association, after learning that those with visual impairments would struggle to know whether their food and drink has gone off without being able to read expiry dates. She soon discovered that her solution, which reacts to the particular conditions in which every specific item is stored, had the potential to help everyone reduce their food waste. Innoscentia, a food-tech company working to counteract this same problem, has developed 'dynamic shelf life labels' for fresh food packaging that change in colour in response to the gases that are produced as a particular item of food begins to go off. 'Use-by' and 'best before dates', that by law must be found on most types of food and drink, are 'worst case scenarios', and 'a guess', according to Pakštaitė. There is no set of regulations stating when a certain food will go off, so individual companies make their own decisions on what dates to print, often adding large safety margins. Whereas use-by dates must be abided by as they are used for foodstuffs like meat and dairy produce that can cause harm if they have gone bad, 'best before' dates simply indicate the estimated date at which the product will begin to lose its freshness, though it is not harmful to consume it after this date.* Research shows that customers are often confused about the difference (and things are even murkier in America where the terms used by manufacturers are not dictated by federal law), leading us to throw away food when it is still perfectly edible, in order to be on the safe side.[28] Conversely, if it is stored in ways counter to instructions, food can go off faster than the use-by date. How often have you accidentally left a carton of milk on the kitchen counter, perhaps between pouring it over your cereal in the morning and your 11 a.m. cup of tea? If they were to be calibrated for milk, Mimica and Innoscentia dynamic labels would be sensitive to the impact of these specific circumstances on the rate at which the carton will expire.

---

* Either a 'use-by' or a 'best before'/ 'best before end' date is legally required on most food and drink in the UK, following the continuation of EU regulation after Brexit. Exceptions include fresh fruit and vegetables (the spoilage of which is noticeable and not of safety concern), wine or other alcoholic drinks over 10 per cent alcohol, vinegar, cooking salt, solid sugar (which naturally resists microbial growth) and baked goods that naturally have a short shelf-life.

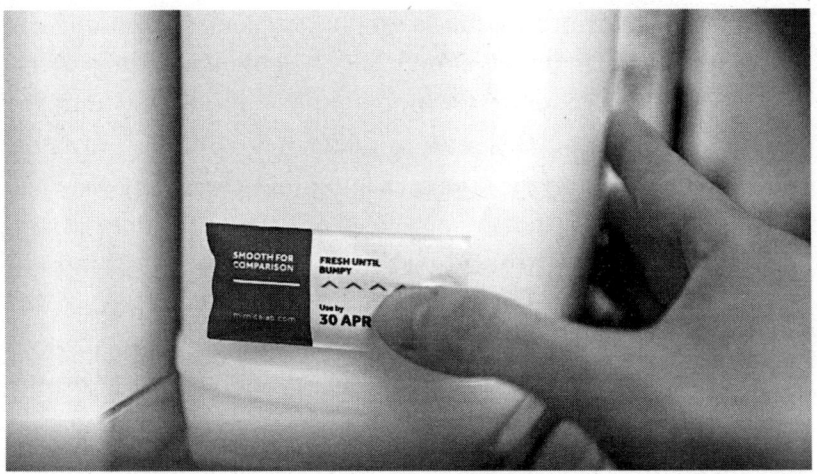

Figure 39. A Mimica Touch label, here applied to a carton of milk, which will change from smooth to bumpy when the specific product is no longer safe to consume.

The expiry date system is a relatively modern invention. According to legend we have the famous American mobster Al Capone to thank, after one of his family members supposedly became sick after drinking gone-off milk and Capone lobbied the milk industry of the 1930s. The reality is more mundane. Marks & Spencer was the first to use 'sell-by' dates in the 1950s as a way of better managing its stock internally, so that staff could move older items to the front of shelves. Sell-by dates were transferred onto the shop floor in the early 1970s, and soon came to be associated – erroneously – with safety standards. In 1979, the EU codified the use of best before and use-by dates into law. Sell-by dates (along with 'display until' dates) continue to cause confusion for customers, who commonly take them to mean that the food has gone off rather than it being for the benefit of the shop, and there have been repeated calls to ban them. For example, in 2009, environment and food secretary Hilary Benn argued that food should only carry a 'use before' date, in a bid to cut down on food waste.[29] In more recent times use-by dates have become the target of reform. In January 2022 Morrisons announced that it would be the first UK supermarket to remove use-by dates from its milk packaging, asking its customers instead to adopt the age-old 'sniff test'. Since milk is pasteurised to kill pathogens the use-by date which indicates safety is not necessary. The best before dates that indicate the period in which a product is at its freshest, will remain, but to work out if the milk has actually gone off, customers will have to rely on its smell.[30] Co-op followed later that year with best before rather than use-by dates on its own-brand yoghurts.[31] Intelligent design products like Mimica Touch and Innoscentia could revolutionise the food labelling business, giving customers confidence beyond their noses to know when food has actually become inedible, and reducing the amount we throw away at home as a result. What's more, food manufacturers that adopt these technologies are more likely to have the confidence to push back their expiry date labels, and supermarkets that stock these products will therefore have a greater window in which to sell them.

Upcycled Leftovers

Finally, innovative ways of making use of the food waste that is produced are ever increasing. I drink my morning coffee from a coffee cup made from leftover coffee beans, and, more dramatically, the technology imagined in the classic 1980s film *Back to the Future* – cars modified to run on food waste – does now exist. In 2017 GENeco, a Bristol-based tech company specialising in 'sustainable solutions for waste', launched the Bio-Bee, the UK's first vehicle to both collect and run on commercial food waste. It collects waste from businesses in Bristol and nearby areas and transports it to an anaerobic digestion plant in Avon, which in turn creates the fuel that gives the Bio-Bee power. Anaerobic digestion is not a new technology – in the early 1900s the science behind it was understood and used to treat sewage, and in the 1990s Germany led in building anaerobic digesters to produce renewable energy[32] – but it is growing as an alternative to landfill. Whereas in many landfill sites the methane created as the food waste is broken down by microorganisms simply goes to harm the environment as it is emitted into the atmosphere,* in anaerobic digesters the release of methane is controlled and collected to generate electricity. Warrens Group, a subsidiary of BioCapital LTD, treat and dispose of waste in the north-east of England. They have built the region's first anaerobic digestion facility – Emerald Biogas – in order to treat food waste and other organic materials from commercial and retail companies and convert it into clean renewable energy and biofuels. Fertiliser, too, made from nitrogen-rich digestate, the remains left over once the gases have been extracted, is sold to local farm owners only in order to prevent further energy expenditure in transportation. It is important, however, that food that could otherwise be fed to humans is not needlessly sent to anaerobic digestion plants, as occurs too often. Tesco, for example, sent 19.898 tonnes of food that could have been consumed by people to anaerobic digestion in 2017–18.[33] Food might also be rescued from anaerobic digestion to be fed to pigs and other animals, which would reduce our reliance on the environmentally draining production of cereal-based feeds, especially that made of soy, the growing of which hugely contributes to the deforestation of the Amazon in South America.

Other companies are harvesting the natural power of insects, as opposed to microorganisms, to break down food waste in a process called bioconversion. As the global population increases, insects are set to provide a greater share of our protein, and insects grown on food waste can be sold on for human

---

* Modern landfills have developed ways of collecting some of the methane and carbon dioxide gas to reuse for electricity.

consumption. They can also be sold as animal feed (although environmentalists concerned with saving energy argue that if the food waste was nutritional enough it should be fed directly to livestock in the first place, in accordance with the food waste hierarchy), while any leftover food waste (along with any dead insect parts or exoskeletons that they have shed) can be sold as fertiliser. Certain insects with high fat contents fed off food waste can even be converted into biodiesel, which again provides an alternative to finite fossil fuels.[34] Controversially, crops like corn have been grown especially to be destroyed by anaerobic digesters as they act as a catalyst to the energy production process, and induce a higher overall output of biofuels.* There is a danger that crops could also be redirected to the bioconversion industry to produce more fuel, rather than relying on genuinely unusable leftovers.

The food waste hierarchy, first established by the EU and now the principal framework for waste management worldwide (see Plate 20), dictates that food waste should be kept in the human food system as far as possible; the production of animal feed, fertilisers and biofuels are lower priorities. As a result, there are a growing number of creative initiatives intended to convert, or 'upcycle', leftovers into new edible products. Tristram Stuart, founder of the food waste charity Feedback, has since gone on to establish ToastAle, a company which uses up leftover bread by fermenting it with malted barley into craft ale. So dedicated is Stuart to the mission that the company feeds profits back into the food waste charity. Beer is made from the same grains found in bread – wheat, barley or rye – and so bakers and brewers have been tied together for thousands of years. Now, bakeries and

---

* While biodiesel is made today from fats like used cooking oil, tallow and those found in general food waste, bioethanol is more often made with crops grown specifically for this purpose. This practice has garnered much criticism and was a leading cause of the world food price crisis as noted in Chapter 6.

sandwich manufacturers (who generally don't use the end slice of the loaf for their sandwiches) can send their surplus bread slices to ToastAle rather than anaerobic digestion factories. As Stuart told me, 'If you want to save the world, you've got to throw a better party than the people destroying it.' Beer made of toast would certainly get the party started![35]

Rubies in the Rubble is another UK company that makes use of discarded food, this time at farm level. Though its strictest rules were relaxed in 2009, the EU's system of classing fruit and vegetables in terms of quality continues to take into account cosmetic characteristics.[*] Supermarkets also have their own aesthetic standards (though, as we've noted, they have got much better at selling 'wonky' ranges and finding alternative uses for these crops). Because suppliers compete for contracts with supermarkets they are in reality forced to adhere to these arbitrary conditions, and will dispose of visually 'imperfect' produce, irrespective of taste, by ploughing it back into the fields, composting, landfilling it or sending it to anaerobic digestion. They also end up growing more food to account for the expected loss to meet their contracts. Carrots are a particularly harshly judged veggie; the FAO estimates that bent, dull in colour, or broken carrots account for 25–30 per cent of the crop, not making it to the supermarket shelves.[36] One study found that the cosmetic grading used on farms for fruits and vegetables results in 6–39 per cent of produce being wasted in relation to total food production in the UK. For strawberries, the authors estimated 10,000 tonnes were wasted as a result of cosmetic standards out of the 102,000 tonnes on average produced in the UK per year. This means 8000 tonnes of avoidable CO_2e emissions in strawberries alone, rising

---

* One potential benefit of Brexit that has been mooted is the possibility of loosening the EU's aesthetic rules. The problem is, if the UK wants to continue to trade with its biggest and nearest trading partner, farms will have to abide by them anyway. Outside of the EU our leaders are of course not able to influence any change in EU cosmetic standards.

to between 60 kilotonnes and 970 kilotonnes when estimating all types of food affected by these criteria.[37] Rubies in the Rubble rescues big, misshapen or blemished fruit and veg directly from farmers to make them into relishes.[38] Environmentally conscious customers might also make use of food delivery services like Oddbox and Wonky Veg Boxes which send rescued food directly to the consumer in a subscription service.

In the dairy industry, leftover whey from yoghurt and cheese production has for centuries been reused as animal feed. Today, whey is made into whey powder, which is used as a protein supplement in a whole range of health-orientated bars and shakes. A project based in Ireland called AgriChemWhey developed a new way of using up the leftover whey permeate, the liquid left over from this process, in 2020. This 'second-generation feedstock – a byproduct of a byproduct', as Dr Bill Morrissey, the leader of the project put it, is transformed into polylactic acid, a type of plastic that can be used in packaging and fabric. 'From a sustainability point of view, it ticks a lot of boxes,' he continued. 'This is very important in terms of climate change.'[39]

Opportunities to upcycle leftovers are as plentiful as there are types of food. Soybean pulp created in the tofu industry and the leftover coffee cherries from the coffee-making process can both be turned into flour. In Los Angeles, which boasts a bountiful juicing culture, Kaitlin Mogentale started Pulp Pantry to tackle the huge amount of pulp that is inevitably produced and is usually sent to farms to feed livestock. Sourcing kale and celery fibre pulp from the huge juicing company Suja, Mogentale makes it into flavourful chips (or crisps, as we would say in the UK).[40] And what about the waste from the meat industry? Icelandic designer Valdís Steinarsdóttir is making food packaging from animal skin and bones, perhaps part of the solution to our modern plastic problem.[41]

Conclusion: Back to the Future of Food Waste

I have described just a small selection of the new and varied technologies being developed to reduce the amount of food waste we produce, and better manage that which we throw away. As this book has shown, innovation has been a key driver in the history of food, whether spearheaded by scientists or housewives. Novel cucumber pickles were imagined in the Tudor kitchen; the tin can fuelled sailors with long-lasting sustenance for perilous overseas voyages; pasteurisation killed off microbes to make liquids like milk safer and last longer; and the fridge revolutionised how we ate and stored our food at home. For good or bad, we have come a long way from the traditional preservation techniques that sustained our ancestors for centuries, when individuals dried, salted, fermented and sugared their harvests in line with the seasonal booms and busts of the agricultural year that shaped everyday life. Now, the global cold chain brings us food from across the world whenever we want it, preserved and packaged long before it reaches our dining tables. Accelerating in the Agricultural Revolution, when sheep, for example, were selectively bred to have less bone and more meat, innovation has also increased the efficiency of the food that we produce. New pesticides, artificial fertilisers and engineered high-yield cereal varieties transformed global agricultural production to feed the booming population of the late twentieth century, while factory farming is the logical result of a quest to produce the greatest amount of profitable meat from livestock. Innovations in public health, what's more, have shielded us from buying decayed or infected food, and from coming into contact with the stinking entrails and rotten vegetables that once made up Tudor muckhills or flowed through busy streets to the detriment of our health. Today in the UK, before it can fester our kitchen waste is removed from our homes in organised rubbish collections and either disposed of in landfill sites or reimagined into energy and other useful products using anaerobic digestion machinery.

However, technologies that increase the amount of safe storage time for a food item, increase its yield or recycle it, do not necessarily reduce the amount of food waste we produce. Instead, fundamentally, retailers and consumers must actually use up the food they buy. In the early 1970s, when freezers were starting to establish themselves as an ordinary feature in our homes, Mary Berry warned her readers that she had had to throw away her stock of frozen blackcurrants when they overran their storage time, left forgotten in the back of the freezer. As she discovered, you should as much as possible only be tempted into buying discounted or bulk foods when it is something that you'll actually eat, else 'you would have done better to keep your purse firmly closed to start with', leading as it often does to a waste of food and money.[42] Likewise, cheap factory-farmed meat might tempt us into buying more and wasting more of it. And the many remarkable modern technologies that help us to recycle waste from the food industry, either by feeding humans and animals with it or by creating other useful by-products, should not overshadow the central aim of reducing waste production and surpluses to start with. Recycling is too often seen as a fast remedy to the spiralling effects of consumerism, but acts perhaps as a plaster not a cure.

Instead, for decades campaigners have urged us to take more time and care in our shopping habits, pausing to check the fridge and our cupboards and writing a shopping list, for example, before we go to the supermarkets. Food waste experts educate us on how to properly organise our fridge to make our food last as long possible, with meat on the lowest shelf, vegetables in the drawer where it is more humid, and leftovers and other more quickly perishable foods on the top shelf. It should always be kept at 5°C or lower. Before we throw away good food, can we learn from the thrifty cooking habits of the past? Perhaps we could return to using up yesterday's joint of meat in sandwiches, frying up stale breadcrumbs, making jams and marmalades from seasonal fruits, transforming milk that has begun to turn by making it into cottage cheese... Just as by-products of the food industry were turned in the past to a whole variety of non-culinary uses at home, from bladder footballs to feather quills, campaigners today tell us to make use of food waste in the house, by transforming old coffee grounds into a face scrub, using banana peel as shoe polish, or scattering eggshells to enrich the garden with calcium.[43]

As we have seen over the course of this book, what we define as 'food waste' – as inedible rather than edible – is not set in stone, but shifts depending on our values and tastes. We've learned how until industrialisation had played out, we were accustomed to eating all the parts of an animal. The blood of a pig was collected and made into a blood pudding, its head became brawn, its trotters were sold as on-the-go snacks, and its stomach was made into sausage casings, its meat cured and stored to last months. It is only as we have become disconnected from the food production processes over the last century that offal products have fallen out of favour in the UK. Most Britons now find offal unappealing or even repellent, an unwelcome reminder that the food we are eating was once a living being with a face. Yet, as Tristram Stuart argued in his 2009 book *Waste*, 'it is surely far more grotesque to show disrespect to the animals we kill by discarding some of their most edible parts', as if their life was taken frivolously and needlessly.[44] If we are to eat meat (and there are, of course, plenty of environmentalists who suggest we should not, given the higher amount of energy and resources that go into producing it), perhaps we should teach ourselves to consume more of the animal once again.

Whether we save or waste food tells us a great deal about our society's wider socio-economic problems and priorities. Imagine if we were to address the issue of food waste today in the light of its devastating environmental impact on the planet, and with the same sense of emergency and urgency that galvanised the government and its citizens during the world wars of the twentieth century. With radio, posters and video campaigns, the people of wartime Britain were taught to make use of every scrap of food in the kitchen, to recognise the value in their used food packages and waste food, and to separate out the different types of household rubbish before it was collected to be reused. This, however, would only be possible today if the government – and the world's governments as a whole – were to believe the expenditure on campaigns, services and innovation to be worth the immediate financial burden and any dip in their political popularity which might come from spending public funds and forcing through radical change. It is money that talks in the modern capitalist age. Whereas in the medieval and early modern period, a universal religious outlook informed consumers of the value of food and castigated wastefulness as sin, in the recent coronavirus pandemic it was primarily financial pressures (alongside the unique physical restrictions of the crisis) that led more of us to reduce our food waste.

If food waste exposes this modern conflict between profit and environmental concerns, it also highlights socio-economic inequality. Just as was true in the past, the modern food system works by some of us having access to more food – and therefore having greater capacity to waste it – than others. In Tudor times this inequality was broadly played out on a local level, with poor peasants dependent upon the scraps thrown out by rich country landlords, who had plenty to go around. Today, the crippling cost of living means that more people than ever are relying on charitable handouts at food banks and other community food redistribution projects, while the richest 5 per cent take home 45 per cent of the nation's total income.[45] Simultaneously, food poverty soars to new heights and obesity has become one of the country's most pressing public health concerns. In times of economic hardship like today, there is a tendency for those in wealth to return to a Victorian (or even Elizabethan) belief in the 'deserving' and 'undeserving' poor, rather than tackling the wider socio-economic and political challenges that keep people in food poverty while others have too much to eat (and waste). The Tory MP Lee Anderson's comments that those who use food banks 'cannot cook properly' and 'cannot budget' remind me of those sanctimonious middle-class Victorians who touted soup made from few ingredients or leftovers as the answer to the poor's hunger problem.[46]

In the modern age, these inequalities are also played out on a global scale. The colonialism and globalisation enforced by the European powers from the sixteenth century means that richer nations eat more food and waste more food at the expense of the global poor. We've seen how in the Second World War, Britain's ability to survive with minimal food shortages on the home front was linked to the devastating famine in colonial India. When our meat comes from animals fed off soy grown on deforested ancestral lands in the Amazon rainforest, or our chocolate from impoverished or even enslaved workers in Burkina Faso and Mali, our experience of plenty and wastefulness often come at the cost of dearth and thrift elsewhere in the world. If we continue to seek out foreign lands to grow our crops as part of a post-Brexit vision of a 'global Britain', will we be further exacerbating resource and food scarcity in these climate-threatened countries? The food that consumers waste in high-income countries, the UK among them, is almost equivalent to all the food produced in sub-Saharan Africa.[47]

The global infrastructures and political choices that put food on the supermarket shelves can seem daunting and insurmountable from the perspective of individuals. Systemic change to tackle food waste and the packaging that comes with it requires governmental intervention and, further, global governmental co-operation. Without the ability to remake the entire food system ourselves, can our desire for change make a difference?

We certainly have the power to change cultural norms, at least on a smaller scale. We have witnessed them changing in our own lifetimes. Some will remember being taught to leave a little food on their plate 'for manners' (perhaps more courteously titled 'Mr Manners', 'Miss Manners' or even 'Lady Manners'), a polite way of suggesting that the food served had satisfied the diner, that dates back at least to the seventeenth century in high-class circles.* Our concern about wasting food means that this custom is basically extinct, and its opposite – leaving a clean plate – is now generally regarded as polite. More and more of us are interested in reducing food waste at home, recycling or composting food scraps and eating sustainably grown foodstuffs.

Importantly, businesses are listening to our changing values. If we express our concern about food waste and environmental policies, supermarkets, cafés and restaurant chains are more likely to take action. I am writing these words while nursing a coffee at a nearby Pret A Manger. Next to me, a big sign proudly advertises the company's 'food waste' consciousness, with any unsold food at the end of the day being donated 'to the hungry [...] and not in the bin'. Clearly, Pret wouldn't do this unless they thought it would be a hit with their customers. Zero waste is indeed now a big industry. The world's first 'zero waste restaurant', Silo, opened its doors in Brighton in 2014. Now based in London, it sources its food directly from farmers, listening to what is seasonally available, uses up all of an animal when it is slaughtered, preserves anything that might go to waste, composts any scraps and even makes use of leftover grains from its own brewery to grow mycelium, a kind of fungus, which becomes modernist lampshades.[48] While in the recent past most were sceptical about the idea of eating upcycled leftovers, however they were altered, it has now

---

* This quote from the seventeenth century might still resonate with us today: 'As our wantons doe at a feast, spare for manners in company but alone cram most greedily'! R. Armin, *Nest of Ninnies* (1608), sig. E.

become 'something of a moral imperative for manufacturers to convert surplus foods into value-added products', according to the food journalist Larissa Zimberoff.[49] Finally, of course, we can vote for political candidates who take the issue of food waste seriously, put pressure on our government to support initiatives that help and call them to account when they don't follow through with meeting food waste and climate change targets.

Even in the absence of governmental support, we can – and are – mobilising as communities to change our relationship to food. New initiatives, like Feedback's Gleaning Network, groups who collectively rescue surplus food from nearby farms, or Bristol's The Children's Kitchen, which teaches nursery school children from food-insecure areas about fresh produce,[50] are bringing us closer to food production. If as a society we learn to love food, not just the quick pleasures of taste and smell that we gain from eating it, but the process of growing, nurturing, pruning, weeding, cooking, preserving and composting, perhaps then we will want to waste less of it. As I hope I've demonstrated throughout *Leftovers*, food is tied up in the rich histories of our culture and society. If we rediscover these connections to what we are eating, and its value to those who have come before us, perhaps we will all waste less.

Whether we will see a meaningful change in our wasteful food culture is left up to the future, dependent as it is on the accumulative power of individual actions alongside governmental regulation. What is clear, however, is that food waste will continue to be an issue deeply tied to our sense of morality, and our relationship to the environment, animals and other human beings. The climate crisis and (to a lesser extent) Brexit and the Covid-19 pandemic mean that we are at a pivotal point in the history of food waste. What history do we want to tell about ourselves to future generations? Ezekias Woodward urged parents in the seventeenth century to teach their children the value of food so that they wasted less of it. 'The parent must remember, and he must remember the childe of it often; That the hungry stomack calls out for bread, bread.'[51] Modern food waste activists continue to beat this same drum, emphasising the importance of bequeathing an appreciation of food to the next generation who will be its guardians, and who will decide what to eat, what to preserve, reuse and what to throw away.

TIMELINE

14th century – In a pattern that was to repeat for at least three centuries, King Edward III ordered the butchers of St Nicholas Shambles outside of London following complaints of entrails and offal in the streets

1540s – London's first sugar refineries opened, though sugar remained an expensive preservative and foodstuff

1558 – Queen Elizabeth I's Religious Settlement re-established Protestantism in England during the Reformation, when religion shaped attitudes to food waste

1562 – A London statute banned citizens from allowing their pigs, goats and poultry to roam freely in the streets

1596 – The Oxfordshire Rebellion, spurred by enclosure policies, a period of poor harvests and raising food prices

1625 – Plague in Cambridge led to orders requiring all food waste be sent to the muckhill outside of town

1640s – First English sugar plantation set up in the colony of Barbados, relying on enforced labour from the transatlantic slave trade, leading to cheaper sugar at home

1660 – 28 November – The Royal Society for the Improvement of Natural Knowledge established, which would support the work of several famous natural philosophers

1665–6 – The Great Plague of London. Miasmas, often created by decaying food, were thought to cause pestilence and disease into the nineteenth century

1670s – Dutchman Antonie van Leeuwenhoek the first to witness bacteria under the microscope, a vital step in our understanding of putrefaction

1671 – John Dwight began the first commercially successful stoneware pottery in England, signifying the widespread use of potting as a new preservation technique[1]

1671 – Act of Common Council in London reinstated a statute banning people from allowing livestock to roam freely in the streets, suggesting that previous orders were unsuccessful; banned throwing animal and food waste into the streets; required residents to gather dirt in tubs to be collected by 'carmen' employed to remove it to a muckhill outside the city

1677 – Samuel Pepys, the secretary to the Admiralty, codified navy rations including many preserved foods

1690s – London orders forbade pig-keeping within the paved part of the city and within 50 yards of any building

1694 – An early recipe for 'pocket soup' appeared in Lady Ann Blencowe's recipe book

18th century – Agricultural Revolution in Britain

— Charles Townshend popularised the Norfolk four-course rotation system
— New inventions included Jethro Tull's seed drill (c. 1701), the Rotherham Swing plough (1730) and the threshing machine (first invented around 1786 by Andrew Meikle)
— Robert Bakewell (1725–95) implemented selective breeding of livestock

1720s – Daniel Defoe's *Tour Thro' the Whole Island of Great Britain* revealed that by this time large-scale pig ownership had grown around industries as they were fed on leftovers

1755 – William Cullen the first to create artificial refrigeration, by evaporating ether in a vacuum, causing it to extract heat from its surroundings

1756 – Mrs Elizabeth Dubois and William Cookworthy awarded contract to supply the Royal Navy with portable soup, a kind of early meat stock cube

1757 – The Sankey Navigation in north-west England built with pressure from the nearby salt industry, marking the start of modern canal building and the faster transport of foodstuffs

1760s – Start of the British Industrial Revolution which would transform the food industry

1765 – Lazzaro Spallanzani published research arguing it was microorganisms in the air that caused putrefaction in food, becoming the first serious rebuttal of the theory of spontaneous generation

1766 – Food riots across England, including the Nottingham Cheese Riot, in response to raised wheat prices following poor harvests

1773 – 16 December – The Boston Tea Party: $18,000 worth of tea thrown into Boston harbour in protest over British rule in America, kick-starting the American Revolution

1780 – John Graefer became the first person to receive a patent for the artificial dehydration of foods

1792 – William Murdoch invented gas lighting, which replaced animal tallow candles and helped fuel the Industrial Revolution

1793 – The Board of Agriculture established to support new inventions in British agriculture

1795 – Hot-air dehydration developed in France, allowing commercial production of dried spaghetti, although this didn't take off until the early twentieth century

1802 – First steam train built, opening up the possibility of transporting food from further afield

1803–15 – The Napoleonic Wars between Britain and France which would spur on advances in food preservation

1810 – Nicholas Appert published *The Art of Preserving Animal and Vegetable Substances* explaining his revolutionary canning method

1813 – Bryan Donkin, John Hall and John Gamble produced the world's first tin cans at their factory in South London

1815 – The Corn Laws passed, protecting British landowners' interests and restricting foreign imports but dramatically increasing bread prices

1817 – An Act for Better Paving, Improving, and Regulating the Streets of the Metropolis aimed to minimise nuisance from pigs and slaughterhouses, and banned the keeping of swine within 40 yards of any street or public place in London

1825 – Opening of the Stockton & Darlington line marked the start of the railway era, allowing food to travel faster over greater distances

1829 – Sir John Ross found Donkin's tin cans in excellent condition from the shipwreck of the *Fury* four years previously

1834 – Jacob Perkins built the first working vapour-compression refrigerator in the world using liquid ammonia, though it did not succeed commercially

1834 – The Poor Law divided poor into 'deserving' and 'undeserving', reflecting Victorian attitudes to food charity

1844 – First transatlantic shipment of ice to be used in ice boxes arrived in England from Wenham Lake in America

1846 – Repeal of the Corn Laws marked a free trade consensus in British politics, to the long-term detriment of British agriculture

1847 – The Town Police Clauses Act legislated across Britain against throwing waste into the street, and in theory forbade swine from being kept 'in or near any street, so as to be a common nuisance'

1847 – Preserved beef in tin cans officially introduced to Royal Navy rations

1848 – Public Health Act formed a Central Board and local Boards of Health with control over slaughterhouses and 'offensive trades', introduced a voluntary municipal system of domestic waste collection and described penalties for those who kept pigsties in their home. An inspector of nuisances empowered to inspect meat and fish to ensure it was fit for consumption

1850s – Devastating cholera outbreak in London blamed on 'miasma', while John Snow's research at the Broad Street Pump suggested it was contaminated water that caused the illness, supporting the germ theory of disease

1851–4 – A report in the *Lancet* alerted the public to the true extent of food adulteration

1852 – Dubbed the 'salt line', the Sandbach and Wheelock branch of the North Staffordshire Railway opened

1852 – Stephan Goldner's faulty tin cans the subject of a Select Committee Inquiry

1855 – Robert Yeates patented a can opener

1858 – The 'Great Stink' in London caused by polluted waterways, much from butchers' waste and other food industries

1860s – French microbiologist Louis Pasteur proved microorganisms cause food spoilage, and invented 'pasteurisation', killing pathogens in liquid via heating. His Swan Neck Flask experiment disproved spontaneous generation and proved germ theory

1861 – Mrs Beeton's hugely influential *Book of Household Management* published in the same year as the final form of Henry Mayhew's *London Labour and the London Poor*

1861 – Thomas Sutcliffe Mort established the first freezing works in the world at Darling Harbour in Sydney, Australia

1868 – John West founded a salmon saltery in Oregon, in which he created an automated tin can-filling machine

1869 – French food chemist Hippolyte Mège-Mouriès invented margarine as a substitute for butter

1874 – Alfred Fryer invented 'the destructor', the first waste incinerator

1875 – Public Health Act established a full set of guidelines for sanitation in towns, consolidating the 1848 Public Health Act and others. Local authorities to remove household rubbish voluntarily, pig-keeping banned in urban areas, slaughterhouses and 'offensive trades' regulated and medical officers empowered to inspect all foods sold

1875 – The Sale of Food and Drugs Act made harmful or fraudulent adulteration of food illegal

1875 – The first shipment of refrigerated meat – from New York to Britain by Timothy Eastman – gained Queen Victoria's royal seal of approval

1877 – The ship *Paraguay* carried the first successful frozen meat cargo between Buenos Aires and Le Havre

1880 – The steamer *Strathleven* successfully transported frozen meat from Australia to England, opening up the trade in the southern hemisphere's surplus meat

1890 – Folding paper carton used in Britain for the first time

1891 – The Public Health (London) Act consolidated the 1875 Public Health Act in London

1901 – Bananas arrived in Britain for the first time in a refrigerated ship, marking the start of a new era of fruit and vegetable importation

1913 – Fred W. Wolf debuted the first domestic refrigerator, the DOMELRE (an acronym of Domestic Electric Refrigerator)

1914–18 – The First World War, leading to a revolution in food waste management

1915 – The Women's Institute, which would play a key role in preserving food throughout both world wars, received government funding to form

1916 – 22 December – The Ministry of Food established in Britain to manage wartime food supplies

1917 – The Women's Land Army formed

1917 – February – Voluntary rationing introduced in Britain

1918 – July – Rationing began for every person in Britain, and ration books were introduced

1918 – German chemist Fritz Haber received the Nobel Prize in Chemistry for inventing the Haber-Bosch process, which fixed nitrogen to make artificial fertilisers possible; Haber's assistant Carl Bosch would receive a Nobel Prize in 1931

1918 – The National Salvage Council formed

1920s – Freon, a CFC refrigerant that replaced the use of dangerous ammonia, invented

1922 – Artificial wax coating to increase shelf life of fresh produce first patented in the US by Ernest Brogden

1922 – A UK law made the pasteurisation of milk a legal requirement

1924 – Clarence Birdseye (who a few years later patented his famous fish fingers) invented the 'quick freeze' or 'flash freeze' method of preserving food

1925 – At the University of Cambridge, Frankin Kidd and Cyril West established the scientific basis for Controlled Atmosphere Storage

1925 – The Public Health (Preservatives etc. in Food) Regulations ruled on what foods could be preserved with chemicals, what chemicals could be used and how they were labelled

1930 – 4 August – World's first supermarket, King Kullen, opened in the United States

1939–45 – The Second World War, requiring top-down food waste strategies

1939 – DDT discovered as an insecticide

1939 – 1 June – The Women's Land Army re-formed

1939 – 8 September – Second Ministry of Food set up

1940 – January – Food rationing introduced, with restrictions on butter, bacon and sugar, followed by meat in March

1940 – The Small Pig Keepers' Council established

1941 – March – The United States' lend-lease programme started, sending food supplies to Britain

1941 – Local councils compelled by the Ministry of Supply to collect salvage

1948 – McDonald's established in San Bernardino, California, heralding the start of the fast food era

1950s – Controlled Atmosphere Storage began to be used on an industrial scale to improve the shelf life of fresh produce

1951 – The UK's first supermarket established – Premier Supermarket in Streatham, South London

1954 – End of war-enforced food waste collections in Britain

1954 – 4 July – With meat off the ration, all rationing ended in Britain

1958 – First commercial use of irradiation as a food preservation method, using electron beams to kill bacteria

1960s – Development of freeze-drying as industrial food preservation technique

1960s–90s – The Green Revolution increased crop yields in the developing world

1962 – Rachel Carson published *Silent Spring*, igniting the New Environmentalist Movement

1962 – The Common Agricultural Policy introduced by the six founding members of the European Union, giving EU farmers subsidies and leading to overproduction in the 1970s

1968 – Astronauts on the *Apollo 8* mission, the first to orbit the Moon, were fed with new thermostabilised 'wetpacks'

1970 – Iceland, selling exclusively frozen food, opened for the first time in Oswestry, Shropshire

1970 – First use of pascalisation – processing under very high pressure to remove spoilage microorganisms and enzymes – to preserve food on a commercial scale

1970 – *The Ecologist* magazine established, offering a critique on the modern food agroindustry

1970 – Norman Borlaug won the Nobel Peace Prize for his work developing Norin 10, a wheat strain that produced higher yields and is more disease resistant

1971 – Greenpeace established

1971 – Friends of the Earth UK established, depositing 1,500 Schweppes glass bottles outside the company's HQ as a protest over disposable bottles

1973 – The UK joined the EU (then known as the European Economic Community, or EEC), starting an era of Europe-wide rules on food waste

1973 – People, which would become the Green Party, established

1974 – October – McDonald's opened its first restaurant in the UK, in Woolwich, South London

1975 – EEC (later EU) established the Waste Framework Directive, which instructs member states to encourage the prevention, recycling and reuse of waste

1979 – The EU codified 'best before' and 'use-by' dates into law

1980s–90s – Outbreak of BSE or 'mad cow disease' in Britain: caused by recycling rendered meat-and-bone meal into animal feed

1987 – Montreal Protocol banned the use of CFCs, which had been used since the 1920s as a refrigerant, on account of the damage they cause to the ozone layer

1988 – UK law banned ruminant protein being fed to ruminants following the BSE outbreak

1990–97 – 'McLibel' trial in Britain

1990s – 1-Methylcyclopropene invented to mimic ethylene, improving the shelf life of fresh produce

1992 – First international gathering of Food Not Bombs hosted in San Francisco, USA

1993 – Kensington and Chelsea in London became first local authority to collect recycling door-to-door

1994 – FareShare first established to feed homeless people

1994 – EU banned use of PAP from mammals in feed of ruminants

1996 – First UK taxes on landfill

1999 – EU established the Landfill Directive, which sets targets on its member states to reduce the amount of biodegradable waste in landfills

2000 – UK government established WRAP

2001 – UK government (then the EU in 2002) banned swill feeding after foot-and-mouth disease caused by the illegal use of untreated waste, which contained infected meat, as pig-feed

2003 – Household Waste Recycling Act obliged all local authorities to provide kerbside collection of at least two types of recycling by 2010

2005 – The Courtauld Commitment established; voluntary targets on cutting waste given to supermarkets

2006 – Bristol became the first UK Core City to introduce separate household food waste recycling collections

2007–8 – World food price crisis coincided with the global financial crisis

2007 – WRAP launched the 'Love Food Hate Waste' campaign

2009 – 16 December – First 'Feeding the 5000' event, in London's Trafalgar Square

2009 – Groceries Supply Code of Practice made it illegal for supermarkets to cancel an order to a supplier

2011 – UK's first biogas plant built, using anaerobic digestion

2012 – Food waste charity Feedback founded

2013 – Groceries Coach Adjudicators Act enforced Groceries Supply Code of Practice

2014 – UK Supermarkets started selling 'wonky veg'

2014 – World's first 'zero waste restaurant', Silo, opened in Brighton

2015 – Leftover-sharing apps Too Good to Go and Olio launched

2015 – IKEA piloted Winnow Solutions, a pioneering AI tool for reducing food waste

2016 – In a referendum the UK voted to leave the EU by a small majority, leaving the UK free to set its own food waste agenda

2018 – An amendment to the EU's Waste Framework Directive included a call to reduce food waste in line with the UN's 2030 Sustainable Development Goal

2018 – 'School strike for climate' movement ignited by Greta Thunberg forced climate change into headlines; food waste is an important part of this conversation

2018 – UK government pledged to reduce 'avoidable' plastic waste by 2042

2020 – 30 January – WHO declared Covid-19 outbreak a global

emergency, marking the start of a period that changed our relationship to food waste

2020 – 31 January – UK officially left the EU, with a transition period until 31 December 2020

2021 – August – EU in part lifted restrictions on PAP, allowing pigs to be fed the leftovers from poultry carcasses

2021 – November – Environment Act promised weekly food waste collections for all UK residents and to enforce separate food waste removal from commercial kitchens

2022 – January – Morrisons announced it would be the first UK supermarket to remove use-by dates from milk

2030 – United Nations Sustainable Development Goal deadline by which to halve food waste as set in 2015

ACKNOWLEDGEMENTS

Like many creative endeavours, this book was first dreamed up during the coronavirus pandemic. I had just graduated from my PhD when the world seemingly shut down and the future became uncertain. My first thanks go to Dan Jones, who encouraged me to use the time to start thinking of a book proposal and who kindly introduced me to his literary agency. I am very grateful to my agent, Rachel Conway, who then believed in the idea and supported me throughout the process. Thanks also go to the book's editor, Richard Milbank, whose enthusiasm and welcome advice helped to shape *Leftovers*.

I have been lucky enough to speak to many fascinating people while researching for this book. Thank you to Tristram Stuart, Josh Eggleton, David Jackson, Solveiga Pakštaitė, Moody Soliman, Phoebe Ruxton, Erik Månsson, Ann Brinkworth and Sue Travers for letting me interview them. Nikki Robertson, Keith McHenry and Anne Baillie kindly spoke to me via email. Thanks are also due to the wonderful community of food and history lovers who follow me as @historyeats over on Instagram, and who provide a constant source of inspiration. In particular thank you to those followers who completed my food survey during the coronavirus pandemic, and to everyone who responded to further research surveys.

I am grateful to Dr Katrina Moseley, with whom I have recently edited a special issue on the history of food waste and sustainability in the *Global Food History* journal, and to those who helped organise and presented at a conference on the topic that we hosted at the University of Cambridge in 2019.

ACKNOWLEDGEMENTS

Thank you to my parents for teaching me to value food and history! And last but not least, I want to thank my wonderful husband Tom, who is my greatest cheerleader and teammate. Without his endless support this book would not have been written.

IMAGE CREDITS

Plate Section Image Credits

1. Library of Congress
2. Patrons' Permanent Fund / National Gallery of Art, Washington
3. © Museo Nacional del Prado
4. Gift of Janice Hammond and Edward Hemmelgarn / The Cleveland Museum of Art
5. © The Merchant's House
6. ©Rijksmuseum
7. © Norfolk Museums
8. Photo © Tate
9. Collection Museum Boijmans Van Beuningen, Rotterdam. Purchased with the support of Rembrandt Association / Photographer: Studio Tromp
10. World History Archive / Alamy Stock Photo
11. Natick Soldier Systems Center Photographic Collection / Digital Commonwealth, Massachusetts Collections Online
12. Waste Not Want Not Plate; Designed by A.W.N. Pugin (English, 1812–1852); Manufactured by Minton and Company (United Kingdom); England; Diam: 33.1 cm; Museum purchase from Decorative Arts Association Acquisition Fund in memory of Dona Guimaraes; 1993-7-1
13. Bovril: "Wherever did I put that Bovril!": beef tea isn't Bovril / Bovril Limited. Source: Wellcome Collection
14. Amoret Tanner / Alamy Stock Photo

15. © British Library Board
16. IWM / Getty Images
17. Mouseion Archives / Alamy Stock Photo
18. Pictorial Press Ltd / Alamy Stock Photo
19. Adam Gasson / Alamy Stock Photo
20. © Parliamentary Copyright
21. Dan Kitwood / Staff / Getty Images
22. © Compton Verney / Bridgeman Images

Internal Image Credits

1. © The Board of Trustees of the Science Museum
2. Maksim / Wikimedia Commons
3. Courtesy of Beinecke Rare Books & Manuscript Library / Yale University Library
4. Left: © The Wilson / Bridgeman Images
   Right: Widener Collection / National Gallery of Art, Washington
5. Halls Holdings House
6. G.S. London / Digitized by Google Books
7. Art Collection 2 / Alamy Stock Photo
8. Internet Archive / University of California Libraries / Wikimedia Commons
9. Heritage Image Partnership Ltd / Alamy Stock Photo
10. © The Board of Trustees of the Science Museum
11. A musk-ox bull on Melville Island, Canada. Mezzotint with engraving by W. Westall, ca 1821, after Lieutenant Beechey. Wellcome Collection. Source: Wellcome Collection
12. © BodminKeep.org.uk
13. © Mary Evans / Kings College London
14. Internet Archive / Wikimedia Commons
15. Chronicle / Alamy Stock Photo
16. © London Metropolitan Archives (City of London) / Heritage-Images

17. Album / Alamy Stock Photo
18. Photo 12 / Alamy Stock Photo
19. GRANGER - Historical Picture Archive / Alamy Stock Photo
20. © Olddesignshop.com
21. © The Board of Trustees of the Science Museum
22. Internet Archive / Wikimedia Commons
23. Hi-Story / Alamy Stock Photo
24. IWM / Getty Images
25. © The Board of Trustees of the Science Museum
26. Chronicle / Alamy Stock Photo
27. Photo: Maidstone Area Arts Partnership / Kentonline.co.uk
28. piemags/archive/military / Alamy Stock Photo
29. How to Make a Compost Heap / Digforvictory.org.uk
30. Library of Congress / Wikimedia Commons
31. National Army Museum, Out of Copyright
32. © The Board of Trustees of the Science Museum
33. International Rice Research Institute
34. PA Images / Alamy Stock Photo
35. Division of Work & Industry, National Museum of American History, Smithsonian Institution
36. Image: James Beck / Bristol Post.co.uk
37. Photo: Sid Chalmers / Fletcher Trust Archives P3000/33
38. With permission from Winnow
39. With permission from Mimica

BIBLIOGRAPHY

Accompts of the Churchwardens of the Paryshe of St Christofer's in London 1572 to 1662, ed. Edwin Freshfield (Rixon and Arnold: London, 1885).

Ackroyd, Peter, *Thames: Sacred River* (Chatto & Windus: London, 2007).

Action Against Hunger, 'World Hunger Facts', https://www.actionagainsthunger.org.uk/why-hunger/world-hunger-facts#:~:text=Globally%2C%20one%20in%20one%20people%20are%20hungry%20or%20undernourished.&text=2.37%20billion%20people%20did%20not,and%20nutritious%20food%20in%202020.

Acton, Eliza, *Modern Cookery for Private Families* (Longmans: London, 1845).

Acts of the Privy Council: Volume 26, 1596–1597, ed. John Roche Dasent (His Majesty's Stationery Office: London, 1902), *British History Online*, http://www.british-history.ac.uk/acts-privy-council.

Adkins, Frankie, 'The Fruitless Saga of the UK's "Pick for Britain" Scheme', *Al Jazeera*, 19 November 2020, https://www.aljazeera.com/features/2020/11/19/pick-for-britain-a-rather-fruitless.

Albala, Ken, *Food in Early Modern Europe* (Greenwood Press: Westport, CT, 2017).

Allen, Gary, *Can It! The Perils and Pleasures of Preserving Food* (Reaktion Books: London, 2016).

Almeroth-Williams, Thomas, *City of Beasts: How Animals Shaped Georgian London* (Manchester University Press: Manchester, 2019).

Althorp Papers, Add MS 75339, British Library, London.

Andrews, Jane, 'Industries in Kent, c. 1500–1640', in Michael Zell (ed.), *Early Modern Kent, 1540-1640* (The Boydell Press/Kent County Council: Woodbridge, 2000), pp. 105–39.

Anon., *A Closet for Ladies and Gentlewomen, or, The Art of Preseruing, Conseruing, and Candying* (F. Kingston for Arthur Johnson: London, 1608).

Anon., *At a Court of Sewers Held at the Guild Hall, London on*

Saterday the Fifth of February in the Year of Our Lord 1652 (Henry Hills for John Bellinger: London, 1653).

Anon., *The Experienced Market Man and Woman, or, Profitable Instructions, to All Masters and Mistrisses of Families, Servants and Others, to Know the Goodness of All Sorts of Provisions* (James Watson: Edinburgh, 1699).

Anon., *The Genteel House-Keepers Pastime, or, The Mode of Carving at the Table Represented in a Pack of Playing Cards* (J. Moxon: London, 1693).

Anon., *The Lady's Companion: or, An Infallible Guide to the Fair Sex* (T. Read: London, 1743).

Anon., *Murray's Modern Cookery Book. Modern Domestic Cookery* (John Murray: London, 1851).

Apeel, 'Comparison to Current Methods for Preservation of Raw Fruits and Vegetables', https://apeelincanada.wordpress.com/2018/11/29/comparison-to-refrigeration-and-controlled-atmosphere-storage/.

Appert, Nicholas, *The Art of Preserving All Kinds of Animal and Vegetable Substances for Several Years* (Black, Parry and Kingsbury: London, 1811).

Ares, Elena and Agnieszka Suchenia, 'Allocations to UK-EU Fisheries Following the UK's Departure from the EU', *House of Commons Library*, 29 November 2021, https://researchbriefings.files.parliament.uk/documents/CDP-2021-0202/CDP-2021-0202.pdf.

Armin, R., *Nest of Ninnies* (T. East for John Deane: London, 1608).

Atkins, Peter (ed.), *Animal Cities: Beastly Urban Histories* (Routledge: London, 2016).

Aubrey, John, *'Brief Lives', Chiefly of Contemporaries, Set Down by John Aubrey, Between the Years 1669 & 1696*, ed. Andrew Clark (Clarendon Press: Oxford, 1898).

Avery, Victoria and Melissa Calaresu (eds.), *Feast & Fast: The Art of Food in Europe 1500-1800* (Philip Wilson: London, 2019).

Bacon, Francis, *A Collection of Apophthegmes New and Old* (Andrew Crooke: London, 1674).

Bailey, Thomas, *Annals of Nottinghamshire. History of the County of Nottingham, Including the Borough*, IV (Simpkin, Marshall and Co.: London, 1852).

Ball, John, *Treatise of Faith Divided into Two Parts. The First Shewing the Nature, the Second, the Life of Faith* (William Stansby for Edward Brewster: London, 1631).

Barnett, Eleanor, 'Fast Food: Eating and Youth Culture in Britain, 1970–2000', *Museum of Youth Culture,* 8 December 2020, https://museumofyouthculture.com/fast-food-youth-culture/.

Barnett, Eleanor, 'Reforming Food and Eating in Protestant England, c. 1560 – c. 1640', *Historical Journal,* 63 (2020), pp. 507–27.

Barnett, Eleanor and Katrina Moseley (eds.), 'Special Issue: Food Waste and Sustainable Eating in Historical Perspective', *Global Food History,* 9 (2023), pp. 221-343.

Baxter, Richard, *A Christian Directory, or, A Summ of Practical Theologie and Cases of Conscience Directing Christians* (Robert White for Nevill Simmons: London, 1673).

Bayly, Lewis, *Practise of Pietie, Directing a Christian How to Walk That He May Please God* (John Hodgets: London, 1613).

Beale, John to Samuel Hartlib, 16 February 1657, The Hartlib Papers 62/24/1A-2B.

Beaty-Pownall, S., *Household Hints* (Horace Cox: London, 1904).

Beckett, Ian F. W., *The Home Front 1914-1918: How Britain Survived the Great War* (Bloomsbury Information: London, 2006).

Beeton, Isabella, *The Book of Household Management* (S. O. Beeton: London, 1861).

Beeton, Isabella, *Mrs Beeton's Book of Household Management* (Ward, Lock and Co.: London, 1907).

Bernard, Richard, *Ruths Recompence, or, A Commentarie Vpon the Booke of Ruth* (Felix Kyngston: London, 1628).

Berry, Mary, *Popular Freezer Cookery* (Octopus Books: London, 1972).

Bloom, Jonty, 'How are Food Supply Networks Coping with Coronavirus?', *BBC News,* 26 March 2020, https://www.bbc.co.uk/news/business-52020648.

Blue Planet II, Episode 7: Our Blue Planet (2017).

'The Booke of the Household of Queene Elizabeth' (1601), in *A Collection of Ordinances and Regulations for the Government of the Royal Household, Made in Divers Reigns* (John Nichols for the Society of Antiquaries: London, 1790), pp. 281–98.

Boorde, Andrew, *Hereafter Foloweth a Compendyous Regyment, or, A Dyetary of Helth* (Robert Wyer for John Gowghe: London, 1542).

Boughey, Joseph and Charles Hadfield, *British Canals: The Standard History* (The History Press: Cheltenham, 2022).

Bown, Stephen, *Scurvy: How a Surgeon, a Mariner, and a Gentleman Solved the Greatest Medical Mystery of the Age of Sail* (Summersdale Publisher Ltd.: Chichester, 2003).

Bradley, Richard, *The Gentleman and Farmer's Guide for the Increase and Improvement of Cattle* (G.S.: London, 1732).

Breverton, Terry, *The Tudor Kitchen: What the Tudors Ate & Drank* (Amberley Publishing: Stroud, 2015).

'Brexit: Why is There a Row Over Fishing Rights?', *BBC News*, 23 December 2021, https://www.bbc.co.uk/news/46401558.

Bright Harbour for Food Standards Agency, 'The Lived Experience of Food Insecurity Under Covid-19', July 2020, https://www.food.gov. uk/sites/default/files/media/document/fsa-food-insecurity-2020_- report-v5.pdf.

Bristol Bites Back Better, 'During the First Lockdown, Food Club Was a Lifesaver as I Was Unable to Go Out to the Shops', 13 March 2021, https://www.goingforgoldbristol.co.uk/during-the-first-lockdown- food-club-was-a-lifesaver-as-i-was-unable-to-go-out-to-the- shops/.

Bristol Bites Back Better, 'Fight Food Waste', https://www. goingforgoldbristol.co.uk/indi-bbbb-area/fight-food-waste/.

Bristol Bites Back Better, 'Surplus Food as a Sustainable Solution', 19 April 2021, https://www.goingforgoldbristol.co.uk/surplus-food-as- a-sustainable-solution/.

Bristol Early Years, 'The Children's Kitchen', https://www. bristolearlyyears.org.uk/health/the-childrens-kitchen/.

'Bristol Going for Gold: Sustainable Food Places Submission', June 2021, https://www.goingforgoldbristol.co.uk/wp-content/ uploads/2021/06/Bristol-Going-for-Gold-HQ-4.pdf.

British Pathé, War Archives, 'Before and After: Easter in Peacetime and Wartime' (1941), 23 Aug 2011, https://www.youtube.com/ watch?v=_lmNyoxujac&ab_channel=WarArchives.

British Pathé, 'Ministry of Information: Bones..Bones..Bones – Save Bones' (24/08/1944), https://www.britishpathe.com/video/bones- bones-bones-save-bones.

British Pathé, 'Queen Visits Salvage Centres 1941' (3/4/1941), https:// www.britishpathe.com/video/queen-visits-salvage-centres.

Browne, Phillis, *The Girl's Own Cookery Book* (The Religious Tract Society: London, 1882).

Buck, W. E. and Mrs Buck, *The 'Little Housewife', or, Domestic Economy for Schools, and Classes in Cookery* (H. Major, and Midland Educational Company: Leicester, 1879).

Bugge, Annechen Bahr, 'Fast Food', in Daniel Thomas Cook, J. Michael Ryan (eds.), *The Wiley Blackwell Encyclopedia of Consumption and*

Consumer Studies (Wiley-Blackwell: Chichester, 2015).

Burnett, John, *England Eats Out: A Social History of Eating Out in England from 1830 to the Present* (Routledge: Abingdon, 2014).

Burnett, John, *Plenty and Want: A Social History of Food in England from 1815 to the Present Day* (Routledge: Abingdon, 1989).

Burton, Robert, *The Anatomy of Melancholy, What It is with All the Kindes, Cavses, Symptomes, Prognostickes, and Severall Cvres of It* (John Lichfield and James Short for Henry Cripps: Oxford, 1621).

Butler, C. T. and Keith McHenry, *Food Not Bombs: How to Feed the Hungry and Build Community* (See Sharp Press: Tucson, AZ, 2000 edn.), https://www.foodnotbombs.net/bookindex.html.

Buttes, Henry, *Dyets Dry Dinner Consisting of Eight Seuerall Courses* (Thomas Creede for William Wood: London, 1599).

Cameron, James Ross (introduced by), *Memory Lane: A Photographic Album of Daily Life in Britain 1930-1953* (J. M. Dent & Sons Ltd.: London, 1980).

Carlson, Cajsa, 'Valdís Steinarsdóttir Turns Animal Skin and Bones into Food Packaging and Vases', *Dezeen*, 27 January 2021, https://www.dezeen.com/2021/01/27/valdis-steinarsdottir-food-packaging-vessels-animal-skin-bones/.

Carr, Esyllt, 'Supermarkets Set Limits on Sale of Cooking Oil', *BBC News*, 22 April 2022, https://www.bbc.co.uk/news/business-61R193141.

Carson, Rachel, *Silent Spring* (Penguin: London, 2000 edn.).

'Certain Profitable and Well Experienced Collections for Making Conserve of Fruits [...] As Also of Surgery, Approved Medicines'. ca. 1650, MS Add 259, Folger Shakespeare Library, Washington DC.

Chadwick, Edwin, *Metropolitan Sewage Committee Proceedings. Parliamentary Papers* 1846, 10:651.

City of Cambridge, *Whereas Divers Disordered People Inhabiting Amongst Us, Not Regarding the Good of this University, and Town of Cambridge* (Cambridge, 1635).

City of London, *The Lawes of the Markette* (John Cawoode: London, 1562).

Clark, Dylan, 'The Raw and the Rotten: Punk Cuisine', *Ethnology*, 43 (2004), pp. 19–31.

Clarkson, L. A., 'The Organization of the English Leather Industry in the Late Sixteenth and Seventeenth Centuries', *The Economic History Review*, 13 (1960), pp. 245–56.

Cleaver, Robert, *A Godlie Form of Householde Governement for the*

Ordering of Priuate Families, According to the Direction of Gods Word (Felix Kingston for Thomas Man: London, 1598).

Clutterbuck, Charlie, *Bittersweet Brexit: The Future of Food, Farming, Land and Labour* (Pluto Press: London, 2017).

Cockayne, Emily, *Hubbub: Filth, Noise & Stench in England: 1600–1750* (Yale University Press: New Haven, CT, 2021).

Codrington, Thomas, *Report on the Destruction of Town Refuse* (Her Majesty's Stationery Office: London, 1888).

Coe, Sarah and Jonathan Finlay, 'Agricultural Act 2020', *House of Commons Library,* Briefing Paper, 3 December 2020, https:// researchbriefings.files.parliament.uk/documents/CBP-8702/CBP-8702.pdf.

Cogan, Thomas, *The Hauen of Health Cheifly Gathered for the Comfort of Students* (Anne Griffin for Roger Ball: London, 1584).

Collingham, Lizzie, *The Taste of War: World War Two and the Battle for Food* (Allen Lane: London, 2011).

Collingham, Lizzie, *The Hungry Empire: How Britain's Quest for Food Shaped the Modern World* (Random House: London, 2018).

Cooper, Tim, 'Challenging the "Refuse Revolution": War, Waste and the Rediscovery of Recycling, 1900–50', *Historical Research,* 81 (2008), pp. 710–31.

Copies of Official Reports and Letters Relative to Donkin, Hall & Gamble's Preserved Provisions (J. Powell: London, 1817).

'Coronavirus: Ancient Mill Resumes Commercial Flour Production', *BBC News,* 22 April 2020, https://www.bbc.co.uk/news/uk-england-dorset-52369075.

'Coronavirus: Nurse's Despair as Panic-Buyers Clear Shelves', *BBC News,* 19 March 2020, https://www.bbc.co.uk/news/av/uk-england-york-north-yorkshire-51966337.

'County: [Warwickshire]. Description of Courts: Manorial Courts', 29 April 1552, SC 2/207/82, National Archives, Kew, London.

Craig, William Marshall, *The Itinerant Traders of London* (Richard Phillips: London, 1804).

Critchell, James Troubridge and Joseph Raymond, *A History of the Frozen Meat Trade: An Account of the Development and Present Day Methods of Preparation, Transport, and Marketing of Frozen and Chilled Meats* (Constable & Company Ltd.: London, 1912).

'Dairy Farmers Forced to Pour Tens of Thousands of Litres of Milk Away Due to HGV Driver Shortage and Rising Costs and Labour Shortages', *Sky News,* 6 October 2021, https://news.sky.com/story/

dairy-farmers-forced-to-pour-tens-of-thousands-of-litres-of-milk-away-due-to-hgv-driver-shortage-and-rising-costs-and-labour-shortages-12427818.

'Damaged and Ugly Fruit to Be Sold by Waitrose in a Bid to Cut Endemic Food Waste', *Daily Mail*, 2 June 2014, https://www.dailymail.co.uk/news/article-2645508/Damaged-ugly-fruit-sold-Waitrose-bid-cut-endemic-food-waste.html.

Davidson, Alan, *The Oxford Companion to Food* (Oxford University Press: Oxford, 2014, 3rd edn.).

Davis, Dorothy, *A History of Shopping* (Routledge: London, 1966).

Daw, Joseph, *A Sketch of the Early History of the Worshipful Company of Butchers of London* (Worshipful Company of Butchers: London, 1869).

Dawson, Mark, *Plenti and Grase: Food and Drink in a Sixteenth-Century Household* (Prospect Books: Totnes, 2009).

Dawson, Thomas, *The Goodhuswifes Iewell Wherein is to Be Found Most Excellent and Rare Deuises for Conceits in Cookerie* (John Wolfe for Edward White: London, 1587).

Dawson, Thomas, *The Second Part of the Good Hus-wiues Iewell Where is to Be Found Most Apt and Readiest Wayes to Distill Many Wholsome and Sweet Waters* (E. Allde for Edward White: London, 1597).

Day, Chris, 'The Great War British Bake Off', The National Archives, 21 September 2015, https://blog.nationalarchives.gov.uk/great-war-british-bake/.

Day, Helen, 'Waste Not Want Not: The Excesses of Gluttony in the Early to Mid-Victorian Period', in Harlan Walker (ed.), *The Fat of the Land: Proceedings of the Oxford Symposium on Food and Cookery 2002* (Footwork: Bristol, 2003), pp. 56–75.

Dean, David, 'Elizabethan Government and Politics', in Robert Tittler and Norman Jones (eds.), *A Companion to Tudor Britain* (Blackwell: Oxford, 2004), pp. 44–60.

Defoe, Daniel, *A Tour Thro' the Whole Island of Great Britain* (G. Strahan: London, 1724–1725), I-II.

Defra, 'United Kingdom Food Security Report 2021', 22 December 2021, https://www.gov.uk/government/statistics/united-kingdom-food-security-report-2021/united-kingdom-food-security-report-2021-theme-2-uk-food-supply-sources#:~:text=Headline-,In%20 2020%2C%20the%20UK%20imported%2046%25%20of%20the%2-0food%20it,%C2%A321.4%20billion%20was%20exported.

Defra Press Office, 'Household Food Waste to Be Collected Separately by 2023 and 50,000 City Trees to Be Planted in Urban Tree Challenge Fund', 10 February 2020, https://deframedia.blog.gov.uk/2020/02/10/household-food-waste-to-be-collected-separately-by-2023-and-50000-city-trees-to-be-planted-in-urban-tree-challenge-fund/.

DEMOS for Food Standards Agency, 'Food in a Pandemic – From Renew Normal: The People's Commission on Life After Covid-19', March 2021, https://demos.co.uk/wp-content/uploads/2021/03/Food-in-a-Pandemic.pdf.

A Descriptive Catalogue of the Naval Manuscripts in the Pepysian Library at Magdalene College, Cambridge: Volume 1, ed. J. R. Tanner (Navy Records Society: London, 1903).

Dickens, Charles, *Bleak House* (Ticknor and Fields: Boston, 1867).

Dickens, Charles, *Oliver Twist, or, The Parish Boy's Progress,* (Project Gutenberg Ebook, 1996).

Dickens, Charles (Jr.), *Dictionary of London: An Unconventional Handbook* (Charles Dickens: London, 1879).

Dolan, Frances E., *Digging the Past: How and Why to Imagine Seventeenth-Century Agriculture* (University of Pennsylvania Press: Philadelphia, PA, 2020).

Downame, John, *A Guide to Godlynesse, or, A Treatise of a Christian Life* (F. Kingston for Philemon Stephens and Christopher Meredith: London, 1622).

Driver, Alistair, 'On-Farm Pig Cull Will Get "Worse and Worse" Unless Summit Delivers Results, Zoe Tells Today Programme', *National Pig Association,* 10 February 2022, http://www.npa-uk.org.uk/On-farm_pig_cull_will_get_worse_and_worse_unless_summit_delivers_results_Zoe_tells_Today_programme.html.

Drury, Joe, 'War on Waste', *Guardian,* 16 May 2001, https://www.theguardian.com/society/2001/may/16/guardiansocietysupplement3.

Durbin, Adam, 'Co-op Supermarket Scraps Yoghurt Use-By Dates in Bid to Cut Food Waste', *BBC News,* 22 April 2022, https://www.bbc.co.uk/news/uk-61184855.

Earle, Rebecca, *Feeding the People: The Politics of the Potato* (Cambridge University Press: Cambridge, 2020).

The Ecologist, Vols. 1:1–6 (1970).

Edwards, Annie Sarah (Oral history), IWM Sound Archive, n. 740 (4/3/1976), reels 1, 2, 7, 8.

EFSA, 'Safety Assessment of the Active Substances Carboxymethylcellulose, Acetylated Distarch Phosphate, Bentonite, Boric Acid and Aluminium Sulfate, for Use in Active Food Contact Materials', *EFSA Journal,* 16 (2018): 5121.

Elkin, Elizabeth, Mai Ngoc Chau and Agnieszka de Sousa, 'Your Food Prices Are at Risk as the World Runs Short of Workers', *Bloomberg,* 2 September 2021, https://www.bloomberg.com/news/features/2021-09-02/food-prices-driven-up-by-global-worker-shortage-brexit.

Elton, G. R., 'Piscatorial Politics in the Early Parliaments of Elizabeth I', in *Studies in Tudor and Stuart Politics and Government* (Cambridge University Press: Cambridge, 2002), IV, pp. 109–30.

Eunomia for WRAP, *Dealing with Food Waste in the UK,* March 2007.

Evans, David, 'Binning, Gifting and Recovery: The Conditions of Disposal in Household Food Consumption', *Society and Space,* 30 (2012), pp. 1123–37.

Evans, David, *Food Waste: Home Consumption, Material Culture and Everyday Life* (Bloomsbury Academic: London, 2014).

Evans, David, Hugh Campbell and Anne Murcott (eds.), *Waste Matters: New Perspectives on Food and Society* (Wiley: Malden, MA, 2013).

Evelyn, John, *Fumifugium, or, The Inconvenience of the Aer and Smoak of London Dissipated* (W. Godbid: London, 1661).

Evelyn, John, *A Philosophical Discourse of Earth Relating to the Culture and Improvement of It for Vegetation, and the Propagation of Plants* (John Martyn: London, 1676).

FABRA UK, 'Environmental Benefits of Rendering', http://www.fabrauk.co.uk/environmental-benefits-of-rendering.

FABRA UK, 'What Are Animal By-Products', FABRA-FA-001, https://static1.squarespace.com/static/5730558e5559862dca6ecc9a/t/605488d54ca0db68f69c04ca/1616152790316/FABRA-FS-001-What+are+Animal+By-products-V1-+210308.pdf.

FABRA UK, 'Rendering – Process and Benefits', FABRA-FS-002, https://static1.squarespace.com/static/5730558e5559862dca6ecc9a/t/6447ced4dfcdde4ca67b9001/1682427605506/FABRA-FS-002-Rendering-Process+and+Benefits-V4-230424%5B1%5D.pdf.

FABRA UK, 'The Circular Economy and Animal By-Products', FABRA-FS-004, https://static1.squarespace.com/static/5730558e5559862dca6ecc9a/t/61c07ea9d0d4552c5cf8 7f13/1640005291410/FABRA-FS-004-The+Circular+Economy+and+Animal+By-products-V2-211206.pdf.

Falconer, William, *A New Universal Dictionary of the Marine,* ed. William Burney (Cadell and W. Davies: London, 1815).

FareShare, 'Budget: FareShare "Deeply Disappointed" No Funding Announced for Food Waste Scheme', 27 October 2021, https://fareshare.org.uk/news-media/press-releases/budget-fareshare-deeply-disappointed/.

FareShare South West, 'Impact Report 2021', March 2021, https://issuu.com/faresharesouthwest/docs/fareshare_impact_report_2021_digital.

Feedback, 'About', *The Gleaning Network,* https://gleaning.feedbackglobal.org/about/.

Feedback, 'Feeding the 5000', https://feedbackglobal.org/campaigns/feeding-the-5000/.

Feedback, 'Gleaning Network', https://feedbackglobal.org/campaigns/gleaning-network/.

Feedback, *The Catering Toolkit: A Guide to Organizing Spectacular and Celebratory Public Events that Tackle Food Waste!,* https://feedbackglobal.org/wp-content/uploads/2016/12/F5K-Catering-toolkit.pdf.

Field, Catherine, '"Many Hands Hands": Writing the Self in Early Modern Women's Recipe Books', in Michelle M. Dowd and Julie A. Eckerle (eds.), *Genre and Women's Life Writing in Early Modern England* (Routledge: Aldershot, 2007), pp. 49–63.

French, Michael and Jim Phillips, *Cheated Not Poisoned?: Food Regulation in the United Kingdom, 1875–1938* (Manchester University Press: Manchester, 2000).

Food and Agriculture Organization of the United Nations, 'Beauty (and Taste!) Are on the Inside', 2018, http://www.fao.org/fao-stories/article/en/c/1100391/.

Food and Agricultural Organization of the United States, 'Global Food Losses and Food Waste', 2011, https://www.fao.org/3/i2697e/i2697e.pdf.

Food Ethics Council, 'Ethics in Our Food Response to Covid-19' (March 2020), https://www.foodethicscouncil.org/ethics-in-our-food-response-to-covid-19/.

Food Standards Agency, 'Covid-19 Consumer Tracker Waves 1-4, 2020', https://www.food.gov.uk/sites/default/files/media/document/covid-19-wave-1-4-report-final-mc.pdf.

Ford Rojas, John-Paul, 'Tory MP Criticised for Saying Food Bank Users Just Need to Learn How to Cook', 11 May 2022, https://news.

sky.com/story/tory-mp-criticised-for-saying-food-bank-users-just-need-to-learn-how-to-cook-12610728.

Forrest, Adam, 'Brexit Checks on EU Food Imports Scrapped, Announces Jacob Rees-Mogg', *The Independent*, 28 April 2022, https://www.independent.co.uk/news/uk/politics/brexit-eu-import-checks-rees-mogg-b2067421.html.

Foy, Karen, *Life in the Victorian Kitchen: Culinary Secrets and Servants' Stories* (Pen & Sword: Barnsley, 2014).

Francis-Devine, Brigid, 'Income Inequality in the UK', House of Commons Library, 30 November 2021.

Franklin, Benjamin, *The Complete Works, in Philosophy, Politics, and Morals, of the Late Dr. Benjamin Franklin*, II (J. Johnson: London, 1806).

'French Ban on Plastic Packaging for Fruit and Vegetables Begins', *BBC News*, 31 December 2021, https://www.bbc.co.uk/news/world-europe-59843697.

Friends of the Earth, 'Our History Campaigning Since 1971', https://friendsoftheearth.uk/who-we-are/our-history.

Fyffes, 'Our Story', https://www.fyffes.com/our-story/our-history/.

Gallagher, Sophie, 'Pick for Britain: How to Get Involved with Campaign Backed by Royals', *The Independent*, 19 May 2020, https://www.independent.co.uk/life-style/pick-britain-how-sign-fruit-vegetable-harvest-a9521976.html.

García-García, R. and S. S. Searle, 'Preservatives: Food Use', in Benjamin Caballero, Paul M. Finglas, Fidel Toldrá (eds.), *Encyclopedia of Food and Health*, 2016, https://www.sciencedirect.com/referencework/9780123849533/encyclopedia-of-food-and-health, pp. 505–9.

Gardner, Marshall B., *A Journey to the Earth's Interior (1920)*, (Global Grey Ebook, 2021).

Gentleman, Amelia, 'Prosecutors Drop Case Against Men Caught Taking Food from Iceland Bins', *The Guardian*, 29 January 2014, https://www.theguardian.com/uk-news/2014/jan/29/prosecutors-drop-case-men-food-iceland-bins.

George V, 'By the King. A Proclamation', 2 May 1917.

Gilligan, Lisa and Kirstin Roberts, 'Managing Separate Food Waste Segregation for All Businesses Set to Come into Force from 2023', *FREETHS*, 17 April 2023, https://www.freeths.co.uk/2023/04/17/mandatory-separate-food-waste-segregation-for-all-businesses-set-to-come-into-force-from-2023/.

Glass, Jessica R. et al., 'Was Frozen Mammoth or Giant Ground Sloth Served for Dinner at the Explorers Club?', *PLOS ONE*, 11 (2016), pp. 1–12.

Glasse, Hannah, *The Art of Cookery, Made Plain and Easy* (London, 1747).

Globe, 30 November 1901.

Glover, Marcus, 'Courtauld: A Model for the World to Follow in the Fight Against Food Waste', *WRAP*, 8 October 2020, https://wrap. org.uk/blog/2020/10/courtauld-model-world-follow-fight-against-food-waste.

Goldstein, David B., 'Manuring Eden: Biological Conversions in *Paradise Lost*', in Hillary Eklund (ed.), *Ground-Work: English Renaissance Literature and Soil Science* (Duguesne University Press: Pittsburgh, PA, 2017), pp. 171–93.

Goodman, Ruth, *How to Be a Tudor: A Dawn-to-Dusk Guide to Tudor Life* (Liveright: New York, 2016).

Goodrich, W. F., *Economic Disposal of Towns' Refuse* (P. S. King & Son: London, 1901).

Gouge, William, *Of Domesticall Duties: Eight Treatises* (William Bladen: London, 1622).

Granger, Thomas, *The Bread of Life, or, Foode of the Regenerate. A Sermon Preached at Botterwike in Holland, Neere Boston, in Lincolnshire* (T. Snodham for Thomas Pavier: London, 1578).

Grantham, Richard B., *A Treatise on Public Slaughter-Houses* (J. Weale: London, 1848).

Grimsby Daily Telegraph, 4 February 1939.

Grose, Francis, *A Classical Dictionary of the Vulgar Tongue* (S. Hooper: London, 1785).

Gunders, Dana, *Waste Free Kitchen Handbook: A Guide to Eating Well and Saving Money by Wasting Less* (Chronicle Books: San Francisco, CA, 2015).

Gwynn, Mary, *Back in Time for Dinner: From Spam to Sushi: How We've Changed the Way We Eat* (Random House: London, 2015).

Haigh, Dorothy Ann (Oral history), IWM Sound Archive, n. 734, (17/02/1976), reel 5.

Hansard, House of Commons Debate: 'Preserved Meats (Navy)', V. 119, 12 February 1852.

Hansard, House of Commons Debate: 'Waste of Food Order (Prosecution)', V. 386, 27 January 1943.

Harrison, William, *The Description of England: The Classic*

Contemporary Account of Tudor Social Life, ed. Georges Edelen (The Folger Shakespeare Library/Dover Publications: New York, 1994).

Heal, Felicity, *Hospitality in Early Modern England* (Clarendon Press: Oxford, 1990).

Henisch, Bridget Ann, *Fast and Feast: Food in Medieval Society* (Pennsylvania State University Press: University Park, PA, 1976).

Hertford County Records: Notes and Extracts from the Sessions Rolls 1581 to 1698, I, ed. W. J. Hardy (C. E. Longmore: Hertford, 1905).

Hindle, Steve, 'Dearth, Fasting and Alms: The Campaign for General Hospitality in Late Elizabethan England', *Past and Present,* 172 (2001), pp. 44–86.

History Extra, 'From William Wordsworth to Extinction Rebellion: A History of Britain's Green Activists', 22 October 2019, https://www.historyextra.com/period/modern/william-wordsworth-extinction-rebellion-history-britains-green-activists-environment-campaigning-activism-environmentalism/.

H.M. Stationery Office, 'A.D. 1780 No. 1275: Preserving Vegetable Substances', *English Patents of Inventions, Specifications,* 3151–3280 (1857).

Hobby, Iris Lilian (Oral history), IWM Sound Archive, n. 18274, reel 2.

Hodgson, Vere, *Few Eggs and No Oranges: The Diaries of Vere Hodgson 1940–45* (Persephone Books: London, 2002).

Holme, Randle, *The Academy of Armory, or, A Storehouse of Armory and Blazon* (Randle Holme: London, 1688).

Hooper, Mary, *Little Dinners: How to Serve Them with Elegance and Economy* (Henry S. King & Co.: London, 1874).

Hounsell, Peter, *London's Rubbish: Two Centuries of Dirt, Dust and Disease in the Metropolis* (Amberley Publishing: Stroud, 2013).

'How FareShare Help Brands Like Tesco Redistribute Surplus Food to Those in Need', *Table Talk Podcast,* 13 April 2021, https://audioboom.com/posts/7842371-how-fareshare-help-brands-like-tesco-redistribute-surplus-food-to-those-in-need.

'How Serious is the Shortage of Lorry Drivers?', *BBC News,* 15 October 2021, https://www.bbc.co.uk/news/57810729.

Hudson, Anne, 'The Mouse in the Pyx: Popular Heresy and the Eucharist', *Trivium,* 26 (1991).

Hughes, Kathryn, *The Short Life and Long Times of Mrs Beeton* (Harper Perennial: London, 2006).

Hughes, Patsy and Antony Seely, 'Making Waste Work', Landfill

Research Paper 96/103, House of Commons (8 November 1996).

Hulme, Alison, *A Brief History of Thrift* (Manchester University Press: Manchester, 2019).

Humphery, Kim, *Shelf Life: Supermarkets and the Changing Cultures of Consumption* (Cambridge University Press: Cambridge, 1998).

Hussain, Nadiya, *Nadiya's Kitchen* (London, 2016), 'Parsnip and Orange Spiced Cake', *Happy Foodie*, https://thehappyfoodie.co.uk/recipes/parsnip-and-orange-spiced-cake.

Imperial War Museum, 'Pig Food: Women's Voluntary Service Collects Salvaged Kitchen Waste, East Barnet, Hertfordshire, England, 1943', https://www.iwm.org.uk/collections/item/object/205200094.

Imperial War Museum, 'Voices of the First World War: Life on the Home Front', https://www.iwm.org.uk/history/voices-of-the-first-world-war-life-on-the-home-front.

IPCC, 'Summary for Policymakers: Special Report on Climate Change and Land', January 2020, https://www.ipcc.ch/srccl/chapter/summary-for-policymakers/.

'Ismaili Muslim Community Supports UK's Vulnerable With Joint 3.5 Tonne Food Donation', 19 November 2020, https://irp-cdn.multiscreensite.com/2fc87325/files/uploaded/FareShare%20Ismaili%20CIVIC%20donation%20FINAL%2018%20November%202020.pdf.

Jack Tar Tuna Fish advert, *Truth*, London, 16 January 1918.

Jack, S., 'What Are Shops Doing About Stockpiling?', BBC News, 22 March 2020, https://www.bbc.co.uk/news/business-51737030 (accessed March 2020).

James, Richard, *A Sermon Concerning the Eucharist Deliuered on Easter Day in Oxford* (Thomas Harper for Robert Allot: London, 1629).

Jarrige, François and Thomas Le Roux, *The Contamination of the Earth: A History of Pollutions in the Industrial Age,* trans. Janice Egan and Michael Egan (The MIT Press: Cambridge, MA, 2020).

Jeswani, Harish K., Gonzalo Figueroa-Torres and Adisa Azapagic, 'The Extent of Food Waste Generation in the UK and its Environmental Impacts', *Sustainable Production and Consumption,* 26 (2021), pp. 532–47.

Johnson, Boris, 'Prime Minister's Statement on Coronavirus (COVID-19): 23 March 2020', *GOV.UK,* https://www.gov.uk/government/speeches/pm-address-to-the-nation-on-coronavirus-23-march-2020.

Jolly, Jasper, '"Use the Sniff Test": Morrisons to Scrap "Use By" Dates from Milk Packaging', *The Guardian*, 9 January 2022, https://www.theguardian.com/business/2022/jan/09/use-the-sniff-test-morrisons-to-scrap-use-by-dates-from-milk-packaging.

King, Peter, 'Gleaners, Farmers and the Failure of Legal Sanctions in England, 1750–1850', *Past and Present*, 123 (1989), pp. 116–50.

Kolberg, Diana I. S. et al., 'Morpholine, Diethanolamine and Triethanolamine Prohibited Additives in the EU', (2012), https://www.eurl-pesticides.eu/library/docs/srm/EPRW_2012_PM_030_Morpholin.pdf.

'Lamentations of Old Father Thames', Harding BII (2049), V1290, Broadside Ballads Online, Bodleian Libraries, Oxford.

The Land Girl, Vols. 1–2 (1940–1942).

Lang, Tim, *Feeding Britain: Our Food Problems and How to Fix Them* (Pelican: London, 2021).

Langert, Bob, *The Battle to Do Good: Inside McDonald's Sustainability Journey* (Emerald Publishing: Bingley, 2019).

Lawrence, Felicity, 'Families Borrowing to Buy Food a Week into UK Lockdown', *The Guardian*, 28 March 2020, https://www.theguardian.com/society/2020/mar/28/families-borrowing-buy-food-week-of-lockdown.

Lawrence, Felicity, 'Millions to Need Food Aid in Days as Virus Exposes UK Supply', *The Guardian*, 27 March 2020, https://www.theguardian.com/world/2020/mar/27/millions-to-need-food-aid-in-days-as-virus-exposes-uk-supply.

Leaver, Thomas, *A Sermon Preached the Thyrd Sondaye in Lente Before the Kynges Maiestie, and His Honorable Counsell* (John Day: London, 1550).

Lewis, C. S., *The Lion, the Witch, and the Wardrobe* (1950, Macmillan: London, 1961 rpt.).

Licence, Tom, *What the Victorians Threw Away* (Oxbow Books: Oxford, 2015).

Lloyd, Paul S., *Food and Identity in England, 1540–1640: Eating to Impress* (Bloomsbury Academic: London, 2015).

London Daily Advertiser, 1747.

London Greenpeace, 'What's Wrong with McDonald's? Everything They Don't Want You to Know', 1986, https://www.mcspotlight.org/case/pretrial/factsheet.html.

Lopez, Annalaura et al., 'Evolution of Food Safety Features and Volatile Profile in White Sturgeon Caviar Treated with Different

Formulations of Salt and Preservatives During a Long-Term Storage Time', *Foods,* 10 (2021): 850.

'Love Food Hate Waste – Broccoli Stalk Soup', *Liverpool Echo,* 5 June 2009, https://www.liverpoolecho.co.uk/news/liverpool-news/love-food-hate-waste---3451592.

Low, George, *A Tour Through the Islands of Orkney and Schetland* (William Peace & Son: Kirkwell, Orkney, 1879).

Mackenzie, Alexander, *Voyages from Montreal, on the River St. Laurence, Through the Continent of North America, to the Frozen and Pacific Ocean; In the Years 1789 and 1793* (T. Cadell, Junn and W. Davies: London, 1801).

MacLachlan, Ian, 'A Bloody Offal Nuisance: The Persistence of Private Slaughter-houses in Nineteenth Century London', *Urban History,* 34 (2007), pp. 227–54.

Malcolmson, Robert and Stephanos Mastoris, *The English Pig: A History* (Hambledon Press: London, 1998).

Malyon, B. H., 'South African Shipping', *Journal of the Royal African Society,* 36 (1937), pp. 438–46.

Markham, Gervase, *Cheap and Good Husbandry for the Well-Ordering of All Beasts and Fowls and for the General Cure of Their Diseases* (W. Wilson for George Sawbridge: London, 1664).

Markham, Gervase, *Countrey Contentments, or, The English Huswife* (John Beale for R. Jackson: London, 1623).

Markham, Gervase, *Markhams Farewell to Husbandry or, The Inriching of All Sorts of Barren and Sterile Grounds in Our Kingdome* (M. Flesher for R. Jackson: London, 1625).

Marshall, Claire, 'Abattoir Labour Shortage Sees Yorkshire Farmer Kill Piglets', *BBC News,* 1 October 2021, https://www.bbc.co.uk/news/science-environment-58749841.

Marshall, William, *The Rural Economy of the Midland Counties; Including the Management of Livestock in Leicestershire and its Environs,* I (G. Nicol: London, 1790).

Mason, Rowena, 'No 10 Corrects "Shop Once a Week" Comment by Shapps', *The Guardian,* 31 March 2020, https://www.theguardian.com/world/2020/mar/31/no-10-slaps-down-shapps-over-shop-once-a-week-comment-coronavirus.

Maxwell, William H., *The Removal and Disposal of Town Refuse* (The Sanitary Publishing Company Ltd.: London, 1898).

May, Robert, *The Accomplisht Cook, or, The Art and Mystery of Cookery* (Nathaniel Brooke: London, 1660).

Mayhew, Henry, 'A Visit to the Cholera Districts of Bermondsey', *The Morning Chronicle*, 24 September 1849.

Mayhew, Henry, *London Labour and the London Poor*, I–III (Griffin, Bohn, and Company: London, 1861–65).

McHenry, Keith, *The Anarchist Cookbook* (See Sharp Press: Tucson, AZ, 2015).

McHenry, Keith, *Hungry for Peace: How you Can Help End Poverty and War with Food Not Bombs* (See Sharp Press: Tucson, AZ, 2012).

McNeill, J. R., *Something New Under the Sun: An Environmental History of the Twentieth-Century World* (WW Norton: New York, 2000).

Meager, Leonard, *The Mystery of Husbandry, or, Arable, Pasture and Wood-land Improved* (Henry Nelme: London, 1697).

Meeker, David L. and C. R. Hamilton, 'An Overview of the Rendering Industry', in David L. Meeker (ed.), *Essential Rendering: All About the Animal By-Products Industry* (National Renderers Association: Alexandria, VA, 2006), pp. 1–16.

'Migrant Farmworkers Whose Harvests Feed Europe Are Blocked at Borders', *New York Times*, 27 March 2020, https://www.nytimes.com/2020/03/27/business/coronavirus-farm-labor-europe.html.

Miles, Lindsay, *The Less Waste No Fuss Kitchen: Simple Steps to Shop, Cook and Eat Sustainably* (Hardie Grant: London, 2020).

Miller, Lettice (Oral History) IWM Sound Archive, n. 14241 (8/1994), reel 1.

Ministry of Agriculture, Fisheries and Food, *Household Food Consumption and Expenditure: 1970 and 1971 with a Review of the Five Years 1966 to 1970: A Report of the National Food Survey Committee* (Her Majesty's Stationery Office: London, 1973).

Ministry of Food, Dig for Victory Leaflet Number 7, 'How to Make a Compost Heap' (1943).

Ministry of Food, Dig for Victory Leaflet Number 16, 'Garden Pests and How to Deal with Them' (1941).

Ministry of Food, 'Making the Most of the Fat Ration' (April 1948).

Ministry of Food, Leaflet Number 27, 'Potatoes' (October 1946).

Ministry of Food, Leaflet Number 11, 'What's Left in the Larder' (December 1946).

Ministry of Food, Leaflet Number 31, 'Cooking for One'.

Ministry of Food, *Potato Pete's Recipe Book* (London, 1943).

Ministry of Food, War Cookery Leaflet Number 4, 'Carrots'.

Ministry of Food, War Cookery Leaflet Number 4, 'Carrots' (July 1943).

Ministry of Food, War Cookery Leaflet Number 11, 'Dried Eggs'.

Moore, Philip, *The Hope of Health Wherein is Conteined a Goodlie Regimente of Life* (John Kyngston: London, 1564).

Murcott, Anne, *Introducing the Sociology of Food and Eating* (Bloomsbury Academic: London, 2019).

Murrell, John, *A New Booke of Cookerie* (John Browne: London, 1615).

Murrell, John, *A Daily Exercise for Ladies and Gentlewomen* (T. Snodham for the widow Helme: London, 1617).

Närvänen, Elina, Nina Mesiranta, Malla Mattila and Anna Heikkinen (eds.), *Food Waste Management: Solving the Wicked Problem* (Palgrave Macmillan: Cham, Switzerland, 2020).

Nash, Thomas, *Nash's Lenten Stuff: Containing the Description and First Procreation and Increase of the Town of Great Yarmouth, in Norfolk,* ed. Charles Hindley (Reeves and Turner: London, 1871).

National Army Museum, 'Women's Land Army Sorting Potatoes, 1944', https://collection.nam.ac.uk/detail.php?q=searchType%3Dsimple%26simpleText%3Dfood%26themeID%3D%26resultsDisplay%3Dlist%26page%3D8&pos=13&total=325&page=8&acc=1986-11-15-47.

National War Savings Committee, 'Mr. Slice O'Bread' (London, 1917).

Neurolabs, 'Technology', https://www.neurolabs.ai/synthetic-data.

Nicola, Maria et al., 'The Socio-economic Implications of the Coronavirus Pandemic (COVID-19): A Review', *International Journal of Surgery,* 78 (June 2020), pp. 185–93.

Norman, Jill, *Eating for Victory: Healthy Home Front Cooking on War Rations* (Michael O'Mara Books Limited: London, 2014), Kindle edn.

Oddy, Derek, *From Plain Fare to Fusion Food: British Diet from the 1890s to the 1990s* (The Boydell Press: Woodbridge, 2003).

Oddy, Derek J., 'From Roast Beef to Chicken Nuggets: How Technology Changed Meat Consumption in Britain in the Twentieth Century', in Derek Oddy and Alain Drouard (eds.), *The Food Industries of Europe in the Nineteenth and Twentieth Centuries* (Routledge: Abingdon, 2016), pp. 231–46.

Olio, 'Our Impact', https://olioex.com/about/our-impact/.

Overton, Mark, *Agricultural Revolution in England: The Transformation of the Agrarian Economy 1500–1850* (Cambridge University Press: Cambridge, 1996).

Page, Christopher, *The Guitar in Tudor England: A Social and Musical History* (Cambridge University Press: Cambridge, 2015).

Papargyropoulou, Effie, Kate Fearnyough, Charlotte Spring and Lucy Antal, 'The Future of Surplus Food Redistribution in the UK: Re-imagining a "Win-Win" Scenario', *Food Policy*, 108 (2022), pp. 1–13.

Parkinson, John, *Paradisi in Sole Paradisus Terrestris. or, A Garden of All Sorts* (Humphrey Lownes and Robert Young: London, 1629).

Parry, William, *Journal of a Voyage for the Discovery of a North-West Passage from the Atlantic to the Pacific* (John Murray: London, 1821).

Parry, William. *Journal of a Third Voyage for the Discovery of a North-West Passage from the Atlantic to the Pacific; Performed in the Years 1824–25* (H. C. Carey and I. Lea: Philadelphia, PA, 1825).

Peavitt, Helen, *Refrigerator: The Story of Cool in the Kitchen* (Reaktion Books: London, 2017).

Pennell, Sara, 'Perfecting Practice? Women, Manuscript Recipes and Knowledge in Early Modern England', in Victoria Burke and Jonathan Gibson (eds.), *Early Modern Women's Manuscript Writing* (Routledge: Aldershot, 2004), pp. 237–58.

Pepys, Samuel, *The Diary of Samuel Pepys*, ed. Henry B. Wheatley (George Bell & Sons: London, 1893), http://www.gutenberg.org/files/4200/4200-h/4200-h.htm - link2H_4_0106.

Percy, Thomas (ed.), *The Regulations and Establishment of the Household of Henry Algernon Percy, the Fifth Earl of Northumberland* (London, 1770).

Perrett, Michelle, 'Coronavirus: Frozen Food Sales Boom by £285m', *Food Manufacture*, 8 July 2020, https://www.foodmanufacture.co.uk/Article/2020/07/08/Coronavirus-Frozen-food-sales-boom-during-lockdown-with-sales-up-285m.

P. L., *By the Directions of the Scriptures [...] To the Tune of Ayme Not Too High* (Francis Grove: London, 1650).

Plat, Sir Hugh, *Certaine Philosophical Preparations of Foode and Beverage for Sea-men, in Their Long Voyage* (H. Lownes: London, 1607).

Plat, Sir Hugh, *Delightes for Ladies to Adorne Their Persons, Tables, Closets and Distillatories* (Peter Short: London, 1602).

Plat, Sir Hugh, *The Iewell House of Art and Nature* (Peter Short: London, 1594).

Plat, Sir Hugh, *Diuerse New Sorts of Soyle*, in *The Iewell House of Art and Nature* (Peter Short: London, 1594).

Plat, Sir Hugh, *Sundrie New and Artificiall Remedies Against Famine* (Peter Short: London, 1596).

Pluskowski, Aleksander (ed.), *Breaking and Shaping Beastly Bodies: Animals as Material Culture in the Middle Ages* (Oxbow: Oxford, 2007).

Porter, Stephen D. et al., 'Avoidable Food Losses and Associated Production-Phase Greenhouse Gas Emissions Arising from Application of Cosmetic Standards to Fresh Fruit and Vegetables in Europe and the UK', *Journal of Cleaner Production,* 201 (2018), pp. 869–78.

Privy Council, *Regulations and Instructions Relating to His Majesty's Service at Sea* (London, 1731).

Quinn, Ian, 'Mandatory Food Waste Reporting Regulation Scrapped', *The Grocer,* 28 July 2023, https://www.thegrocer.co.uk/food-waste/mandatory-food-waste-reporting-regulation-scrapped/681656.article.

Randall, Adrian, *Riotous Assemblies: Popular Protest in Hanoverian England* (Oxford University Press: Oxford, 2006).

Rappaport, Steve, *Worlds Within Worlds: Structures of Life in Sixteenth-Century London* (Cambridge University Press: Cambridge, 1989).

Rathbone, Hugh R., Letter to the *Liverpool Echo,* 12 January 1918, The National Archives, Kew, London, MAF 60/251b.

Receipt Book of Sarah Longe ca. 1610, MS Add 444, Folger Shakespeare Library, Washington DC.

Redfern, Aimi, '"100,000 Litres of Milk Wasted" as Lorry Driver Shortage Hits Dairy Farmers', *StokeonTrentLive,* 10 September 2021, https://www.stokesentinel.co.uk/news/stoke-on-trent-news/100000-litres-milk-wasted-lorry-5893962.

ReFED 'Investment Tracker', https://refed.org/investment-tracker/ (accessed April 2022).

Reynolds, Christian, Tammara Soma, Charlotte Spring and Jordon Lazell (eds.), *Routledge Handbook of Food Waste* (Routledge: Abingdon, 2020).

'Richard II: January 1393', in *Parliament Rolls of Medieval England,* ed. Chris Given-Wilson, et al. (Woodbridge, 2005), *British History Online,* (accessed April 2022). https://www.british-history.ac.uk/no-series/parliament-rolls-medieval/january-1393.

Riley, Jonathan, 'Crisis in Pig Sector Deepens as Welfare Cull Tops 30,000', *Farmers Weekly,* 16 December 2021, https://www.fwi.co.uk/livestock/crisis-in-pig-sector-deepens-as-welfare-cull-tops-30000.

Riley, Mark, 'From Salvage to Recycling – New Agendas or Same Old Rubbish?', *Area,* 40 (2008), pp. 79–89.

Robertson, Gordon L., *Food Packaging: Principles and Practice,* (CRC Press: Boca Raton, FL, 2006).

Robinson, Doris (Oral history), IWM Sound Archive, n. 12582 (1/6/1992), reel 1.

Rowlands, Samuel, *A Sacred Memorie of the Miracles Wrought by Our Lord and Sauiour Iesus Christ* (Bernard Alsop: London, 1618).

The Royal Family Channel, 'Prince Charles Urging People to "Pick for Britain"' (19 May 2020), https://www.youtube.com/ watch?v=taHhUoxvBL8&ab_channel=TheRoyalFamilyChannel.

Rubies in the Rubble, 'Our Story', https://rubiesintherubble.com/ pages/our-story.

Rumbold, Margaret Mary (Oral history), IWM Sound Archive, n. 8856 (31/05/1985), reels 1-4.

Sabine, Ernest L., 'Butchering in Mediaeval London', *Speculum,* 8 (1993), pp. 335–53.

Sabine, Ernest L., 'City Cleaning in Mediaeval London', *Speculum,* 12 (1937), pp. 19–43.

Saladino, Dan, *Eating to Extinction: The World's Rarest Foods and Why We Need to Save Them* (Jonathan Cape: London, 2021).

Schraer, Rachel, 'EU Fishing Rules: Did the UK Throw Away a Million Tonnes of Fish?', *BBC News,* 11 February 2020, https://www.bbc. co.uk/news/uk-51415240.

Scott, Caroline, *Holding the Home Front: The Women's Land Army in the First World War* (Pen & Sword: Barnsley, 2017).

Scunthorpe Evening Telegraph, 4 February 1939.

Second World War Experience Centre, 'Events in North Africa – June 1942', https://war-experience.org/events/events-in-north-africa-june-1942/.

Sgroi, R. C. L., 'Piscatorial Politics Revisited: The Language of Economic Debate and the Evolution of Fishing Policy in Elizabethan England', *Albion: A Quarterly Journal Concerned with British Studies,* 35 (2003), pp. 1–24.

Shakespeare, William, *The Merry Wives of Windsor* (1602).

Shakespeare, William, *Much Ado About Nothing* (1612).

Shephard, Sue, *Pickled, Potted & Canned: How the Art and Science of Food Preserving Changed Civilisation* (Headline Books: London, 2000).

Shewring, Adam, *The Plain-Dealing Poulterer* (C. Brome: London, 1687).

Silo, 'Becoming Zero Waste', https://silolondon.com/story/.

Sim, Alison, *The Tudor Housewife* (Sutton: Stroud, 2005).

Sloane MS 2244, British Library, London.

Smiles, Samuel, *Self-Help: With Illustrations of Character and Conduct* (John Murray: London, 1859).

Smith, Malcolm C. et al., 'Biomedical Results of Apollo', *NASA History Division*, https://history.nasa.gov/SP-368/s6ch1.htm.

Smith, Roff, 'How Reducing Food Waste Could Ease Climate Change', *National Geographic*, 23 January 2015, https://www.nationalgeographic.com/science/article/150122-food-waste-climate-change-hunger.

Smithers, Rebecca, 'Morrisons Gives Food Banks £10m During Coronavirus Outbreak', *The Guardian*, 30 March 2020, https://www.theguardian.com/world/2020/mar/30/morrisons-gives-food-banks-10m-during-coronavirus-outbreak.

Smith-Howard, Kendra, *Pure and Modern Milk: An Environmental History Since 1900* (Oxford University Press: Oxford, 2017).

Social Metrics Commission, *Measuring Poverty* (July 2019).

Spanner Films, 'McLibel: Two People Who Wouldn't Say Sorry', (2005).

Speed, Adolphus, *Adam out of Eden or, An Abstract of Divers Excellent Experiments Touching the Advancement of Husbandry* (Henry Brome: London, 1658).

Spurling, Hilary (ed.), *Elinor Fettiplace's Receipt Book* (Penguin: Harmondsworth, 1987).

Stedman, John Gabriel, *Narrative, of a Five Years' Expedition; Against the Revolted Negroes of Surinam: in Guiana, on the Wild Coast of South America; From the Year 1772, to 1777* (J. Johnson: London, 1796).

Stedman-Jones, Gareth, *Outcast London: A Study in the Relationship Between Classes in Victorian Society* (Verso: London, 2013).

Steffen, Andrea D., 'New Processes Turn Dairy Waste into Bioplastics and Fertilizers', *Intelligent Living*, 23 July 2020, https://www.intelligentliving.co/process-dairy-waste/.

Stokes, Raymond G., Roman Köster and Stephen C. Sambrook, *The Business of Waste: Great Britain and Germany, 1945 to the Present* (Cambridge University Press: Cambridge, 2013).

Stokoe, Joan, 'Banana Sandwiches' by Northumberland County Libraries, WW2 People's War https://www.bbc.co.uk/history/ww2peopleswar/stories/03/a2734003.shtml.

Stone, Richard, *Mammoth: The Resurrection of an Ice Age Giant* (Fourth Estate: London, 2002).

Stow, John, *A Survey of London. Reprinted from the Text of 1603,* ed. C. L. Kingsford (Clarendon: Oxford, 1908).

Stuart, Tristram, *Waste: Uncovering the Global Food Scandal* (Penguin: London, 2009).

Summers, Julie, *Jambusters: The Remarkable Story Which Has Inspired the ITV Drama Home Fires* (Simon & Schuster: London, 2013).

Sutton, John Frost and Henry Field, *The Date-Book of Remarkable & Memorable Events Connected with Nottingham and its Neighbourhood 1750–1879* (The Proprietor: Nottingham, 1880).

Swinburne, Layinka, 'Dancing with the Mermaids: Ship's Biscuit and Portable Soup', in Harlan Walker (ed.), *Food on the Move: Proceedings of the Oxford Symposium on Food and Cookery* (Prospect Books: Totnes, 1997), pp. 309–20.

Thomson, Ben, 'Changes to Food Waste Collection and Recycling: What Actions do Councils Need to Take?', *LocalGov,* 17 May 2022, https://www.localgov.co.uk/Changes-to-food-waste-collection-and-recycling-What-actions-do-councils-need-to-take-/54208.

Thorsheim, Peter, *Inventing Pollution: Coal, Smoke, and Culture in Britain Since 1800* (Ohio University Press: Athens, OH, 2006).

Thorsheim, Peter, *Waste into Weapons: Recycling in Britain During the Second World War* (Cambridge University Press: Cambridge, 2015).

Tolmachoff, I. P., 'The Carcasses of the Mammoth and Rhinoceros Found in the Frozen Ground of Siberia', *Transactions of the American Philosophical Society,* 23 (1929), pp. i–74b.

Trigge, Francis, *To the Kings Most Excellent Maiestie. The Humble Petition of Two Sisters; The Church and Commonwealth* (George Bishop: London, 1604).

The Trussell Trust, 'Church Support', https://www.trusselltrust.org/get-involved/church-support/.

The Trussell Trust, 'What We Do: Our Vision is for a UK Without the Need for Food Banks', https://www.trusselltrust.org/what-we-do/.

Tryon, Thomas, *The Way to Health, Long Life and Happiness, or, A Discourse of Temperance* (H. C. for D. Newman: London, 1691).

Tryon, Thomas, *Tryon's Letters Upon Several Occasions* (George Conyers and Elizabeth Harris: London, 1700).

Tudor Royal Proclamations, Volume 1: The Early Tudors (1485–1553), eds. Paul L. Hughes and James F. Larkin (Yale University Press: New Haven, CT, 1964).

Tusser, Thomas, *Fiue Hundreth Points of Good Husbandry Vnited to as Many of Good Huswiferie* (Richard Tottill: London, 1573).

Uba, Estelle, 'Ministers Criticised for Scrapping New Food Waste Laws for England', *The Guardian*, 17 August 2023, https://www.theguardian.com/environment/2023/aug/17/ministers-criticised-for-scrapping-new-food-waste-laws-for-england.

UK Government, 'Press Release: New Plans Unveiled to Boost Recycling', 7 May 2021, https://www.gov.uk/government/news/new-plans-unveiled-to-boost-recycling.

UK Government, 'Statutory Guidance: Landing Obligation General Requirements', *GOV.UK*, 21 July 2021, https://www.gov.uk/government/publications/technical-conservation-and-landing-obligation-rules-and-regulations-2021/landing-obligation-general-requirements_v2023.pdf.

UK Government, 'Statutory Guidance: Landing Obligation General Requirements', GOV.UK, Updated April 2023, https://assets.publishing.service.gov.uk/media/643816a9773a8a0013ab2c0c/Landing_obligation_general_requirements_v2023.pdf.

Uliarte, Ernesto Martin et al., 'A Large-Scale Composting Experimentation Using Grape Marcs and Lees Residues from Mendoza Wine Alcohol Industry', Conference paper XXth Giesco International Meeting (2017).

United Nations Environment Programme, *Food Waste Index Report 2021* (Nairobi, 2021).

University of Cambridge, *Articles and Orders Agreed Vpon by the Right Worshipfull John Mansel Doctor of Divinite and Vicechancellor of the Vniversitie of Cambridge* (Cambridge, 1625).

VintageBritishComedy, 'Elsie & Doris Waters - the Kitchen Front (20/12/1941), 4 June 2012, https://www.youtube.com/watch?v=snauq-Yglhs.

W. A., *A Booke of Cookrye. Very Necessary for All Such as Delight Therein* (Edward Allde: London, 1587).

Walford, Nigel, 'The Extent and Impact of the 1940 and 1941 "Plough-up" Campaigns on Farming Across the South Downs, England', *Journal of Rural Studies*, 32 (2013), pp. 38–49.

Wall, Wendy, 'Just a Spoonful of Sugar: Syrup and Domesticity in Early Modern England', *Modern Philology*, 104 (2006), pp. 149–72.

Walter, John, *Crowds and Popular Politics in Early Modern England* (Manchester University Press: Manchester, 2010).

Wasteless, https://www.wasteless.com/.

Watton, Cherish, 'Women's Land Army & Timber Corp', https://www.

womenslandarmy.co.uk/first-world-war-womens-land-army/
recruitment/.

Weinreb, Alice Autumn, *Modern Hungers: Food and Power in
Twentieth-century Germany* (Oxford University Press: Oxford, 2017).

White, Matthew, 'The Industrial Revolution', *British Library*, 14
October 2009, https://www.bl.uk/georgian-britain/articles/the-
industrial-revolution.

Wilson, C. Anne (ed.), *Waste Not, Want Not: Food Preservation
from Early Times to the Present Day* (Edinburgh University Press:
Edinburgh, 1991).

Winnow, 'IKEA and Winnow Are Building the Kitchen of the Future',
https://info.winnowsolutions.com/ikea-and-winnow-are-building-
the-kitchen-of-the-future.

Wolfe, Heather, 'Early Modern Straws; or, Quills are Not Just for
Writing', *The Collation*, 2 October 2019, https://collation.folger.
edu/2019/10/quills-not-just-for-writing/.

Women's Institute, 'History', 'The WI: Inspiring Women', https://www.
thewi.org.uk/about-us/history-of-the-wi.

Women's Land Army LAAS Handbook, *c.* 1917.

Wood, Andy, *The Memory of the People: Custom and Popular Senses
of the Past in Early Modern England* (Cambridge University Press:
Cambridge, 2013).

Woodward, Ezekias, *A Childes Patrimony. Laid Out Upon the Good
Nurture, or Tilling Over the Whole Man* (I. Legat: London, 1640).

Woolley, Hannah (falsely attributed), *The Accomplish'd Lady's Delight
in Preserving, Physick, Beautifying, and Cookery* (B. Harris: London,
1675).

Woolley, Hannah, *The Ladies Directory in Choice Experiments &
Curiosities of Preserving in Jellies, and Candying Both Fruits &
Flowers* (T. M. for Peter Dring: London, 1662).

Woolley, Hannah, *The Queen-like Closet, or, Rich Cabinet Stored with
All Manner of Rare Receipts for Preserving, Candying & Cookery* (R.
Lowndes: London, 1670).

WRAP, 'Food Surplus and Waste in the UK – Key Facts', October 2021,
https://wrap.org.uk/sites/default/files/2021-10/food-%20surplus-
and-%20waste-in-the-%20uk-key-facts-oct-21.pdf.

WRAP, 'Food Surplus and Waste in the UK – Key Facts', December
2022, https://wrap.org.uk/sites/default/files/2023-01/Food%20
Surplus%20and%20Waste%20in%20the%20UK%20Key%20
Facts%20December%202022.pdf

WRAP, 'History of the Courtauld Commitment', https://wrap.org.
uk/taking-action/food-drink/initiatives/courtauld-commitment/
history-courtauld-commitment.

WRAP, 'Let's Keep Crushing It', *Love Food Hate Waste,* https://www.
lovefoodhatewaste.com/keepcrushingit (accessed June 2022).

WRAP, 'Life Under Covid-19: Food Waste Attitudes and Behaviours
in 2020', November 2020 https://wrap.org.uk/resources/report/life-
under-covid-19-food-waste-attitudes-and-behaviours-2020.

WRAP, 'Love Food Hate Waste', https://www.lovefoodhatewaste.com/
recipes.

WRAP, 'Love Food Hate Waste Campaigns', https://wrap.org.uk/
taking-action/citizen-behaviour-change/love-food-hate-waste/key-
campaigns (accessed June 2022).

WRAP, 'Returning to Normality After Covid-19: Food Waste Attitudes
and Behaviours in 2021', August 2021, https://wrap.org.uk/sites/
default/files/2021-08/food-trends-report-august-2021.pdf.

WRAP, 'UK Household Food Waste Tracking Survey Winter 2021:
Behaviours, Attitudes, and Awareness', February 2022, https://wrap.
org.uk/sites/default/files/2022-03/WRAP-UK-household-food-
waste-Winter-2021-Behaviours-attitudes-and-awareness.pdf.

Wright, Mic, 'Has "Best Before" Reached its Sell-By Date?', *The
Guardian,* 16 January 2009, https://www.theguardian.com/
lifeandstyle/wordofmouth/2009/jun/16/food-waste-best-before-
dates.

Wynter, A., *Our Social Bees, or, Pictures of Town & Country Life*
(Robert Hardwicke: London, 1861).

Zero Waste Europe, 'The Story of Too Good To Go', (2020), https://
zerowasteeurope.eu/wp-content/uploads/2020/01/zero_waste_
europe_CS7_CP_TooGoodToGo_en.pdf.

Zimberoff, Larissa, *Technically Food: Inside Silicon Valley's Mission to
Change What We Eat* (Abrams Press: New York, 2021).

Zimring, Carl A. and William L. Rathje (eds.), *Encyclopedia of
Consumption and Waste: The Social Science of Garbage* (SAGE: Los
Angeles, CA, 2012).

NOTES

Introduction

1. As described in Isabella Beeton, *Mrs Beeton's Book of Household Management* (Ward, Lock, and Co.: London, 1907), p. 297.
2. Food and Agriculture Organization of the United Nations, 'Beauty (and Taste!) Are on the Inside', 2018, http://www.fao.org/fao-stories/article/en/c/1100391/.
3. Tristram Stuart, *Waste: Uncovering the Global Food Scandal* (Penguin: London, 2009), p. 140.
4. Ibid., pp. 95–6.
5. Dana Gunders, *Waste Free Kitchen Handbook: A Guide to Eating Well and Saving Money by Wasting Less* (Chronicle Books: San Francisco, CA, 2015), p. 19.
6. Helen Day, 'Waste Not Want Not: The Excesses of Gluttony in the Early to Mid-Victorian Period', in Harlan Walker (ed.), *The Fat of the Land: Proceedings of the Oxford Symposium on Food and Cookery 2002* (Footwork: Bristol, 2003), p. 64.
7. Roff Smith, 'How Reducing Food Waste Could Ease Climate Change', *National Geographic,* 23 January 2015, https://www.nationalgeographic. com/science/article/150122-food-waste-climate-change-hunger; WRAP, 'Food Surplus and Waste in the UK- Key Facts', October 2021, https:// wrap.org.uk/sites/default/files/2021-10/food-%20surplus-and-%20waste-in-the-%20uk-key-facts-oct-21.pdf, pp. 13–14.
8. Stats calculated by the UN's Food and Agriculture Organisation, as quoted in 'Bristol Going for Gold: Sustainable Food Places Submission', June 2021 https://www.goingforgoldbristol.co.uk/wp-content/uploads/2021/06/Bristol-Going-for-Gold-HQ-4.pdf, p. 42; Elina Närvänen, Nina Mesiranta, Malla Mattila, and Anna Heikkinen, 'Introduction: A Framework for Managing Food Waste', in ibid. (eds.), *Food Waste Management: Solving the Wicked Problem* (Palgrave Macmillan: Cham, Switzerland, 2020), p. 1.
9. WRAP, 'Let's Keep Crushing It', *Love Food Hate Waste,* https://www. lovefoodhatewaste.com/keepcrushingit (accessed June 2022).

10. Thomas Bailey, *Annals of Nottinghamshire. History of the County of Nottingham, Including the Borough,* IV (Simpkin, Marshall, and Co.: London, 1852), pp. 2–3; John Frost Sutton and Henry Field, *The Date-Book of Remarkable & Memorable Events Connected with Nottingham and its Neighbourhood 1750–1879* (The Proprietor: Nottingham, 1880), pp. 69–70.

11. Adrian Randall, *Riotous Assemblies: Popular Protest in Hanoverian England* (Oxford University Press: Oxford, 2006), pp. 92–5.

12. See, for example, Richard James, *A Sermon Concerning the Eucharist Deliuered on Easter Day in Oxford* (Thomas Harper for Robert Allot: London, 1629), p. 10: 'if by chance we see it nibled with mice, putrified with wormes, or deuoured by dogs'; Anne Hudson, 'The Mouse in the Pyx: Popular Heresy and the Eucharist', *Trivium,* 26 (1991), p. 43.

13. Leonard Meager, *The Mystery of Husbandry, or, Arable, Pasture and Wood-land Improved* (Henry Nelme: London, 1697), p. 75.

14. L. Clarke, 'Gassing Rats', *The Land Girl,* Vol. 2, n. 12 (March 1942), p. 8.

15. R. García-García and S. S. Searle, 'Preservatives: Food Use', in Benjamin Caballero, Paul M. Finglas, Fidel Toldrá (eds.), *Encyclopedia of Food and Health,* 2016, https://www.sciencedirect.com/referencework/9780123849533/encyclopedia-of-food-and-health, pp. 505–9.

16. Gary Allen, *Can It!: The Perils and Pleasures of Preserving Foods* (Reaktion Books: London, 2016), p. 61.

17. *Blue Planet II,* Episode 7: Our Blue Planet (2017).

18. 'French Ban on Plastic Packaging for Fruit and Vegetables Begins', *BBC News,* 31 December 2021, https://www.bbc.co.uk/news/world-europe-59843697.

19. Stats from Environmental Investigation Agency for Greenpeace UK, quoted in Tim Lang, *Feeding Britain: Our Food Problems and How to Fix Them* (Pelican: London, 2021), p. 135.

20. Richard Stone, *Mammoth: The Resurrection of an Ice Age Giant* (Fourth Estate: London, 2002), p. 34.

21. Jessica R. Glass et al., 'Was Frozen Mammoth or Giant Ground Sloth Served for Dinner at the Explorers Club?', *PLOS ONE,* 11 (2016), p. 7. Claims that the meat of the mammoth discovered in the Beresova River in Siberia in 1901 was eaten at a banquet are sourced from the hollow-earth conspirator Marshall B. Gardner's *A Journey to the Earth's Interior (1920),* (Global Grey Ebook, 2021), p. 93. They are rebuked by the Russian geologist I. P. Tolmachoff in his 'The Carcasses of the Mammoth and Rhinoceros Found in the Frozen Ground of Siberia', *Transactions of the American Philosophical Society,* 23 (1929), p. 60.

22. Francis Bacon, *A Collection of Apophthegmes New and Old* (Andrew Crooke: London, 1674), p. 15. The tale is also retold in Isabella Beeton, *The Book of Household Management* (S. O. Beeton: London, 1861), p. 379.

23. Alexander Mackenzie, *Voyages from Montreal, on the River St. Laurence, Through the Continent of North America, to the Frozen and Pacific Ocean; In the Years 1789 and 1793* (T. Cadell, Junn and W. Davies: London, 1801), e.g. pp. 241, 285–6, 343.

24. Alan Davidson, *The Oxford Companion to Food* (Oxford University Press: Oxford, 2014, 3rd edn.), p. 107.

25. For example, W. E. Buck and Mrs Buck, *The 'Little Housewife', or, Domestic Economy for Schools, and Classes in Cookery* (H. Major, and Midland Educational Company: Leicester, 1879), pp. 57–8; Mary Hooper, *Little Dinners: How to Serve Them with Elegance and Economy* (Henry S. King & Co.: London, 1874), p. 153.

26. Allen, *Can It!*, p. 185.

27. Stuart, *Waste*, p. 44. The Marmite Food Extract Company was formed in 1902 and used up the by-products of the nearby Bass Brewery.

28. The Oxford English Dictionary gives 1642 as its first printed example of this earlier phrase. *OED* 'waste not'.

29. Tom Licence, *What the Victorians Threw Away* (Oxbow Books: Oxford, 2015).

30. Jill M. Church, 'Archaeology of Garbage', in Carl A. Zimring and William L. Rathje (eds.), *Encyclopedia of Consumption and Waste: The Social Science of Garbage*, I (SAGE: Los Angeles, CA, 2012), p. 33.

31. This book is indebted to the work of sociologists who have pioneered in the academic study of food waste. See, for example, David Evans, Hugh Campbell and Anne Murcott (eds.), *Waste Matters: New Perspectives on Food and Society* (Wiley: Malden, MA, 2013); David Evans, *Food Waste: Home Consumption, Material Culture and Everyday Life* (Bloomsbury Academic: London, 2014); Christian Reynolds, Tammara Soma, Charlotte Spring and Jordon Lazell (eds.), *Routledge Handbook of Food Waste* (Routledge: Abingdon, 2020).

Chapter 1: Pickles and Piety

1. Robert May, *The Accomplisht Cook, or, The Art and Mystery of Cookery* (Nathaniel Brooke: London, 1660), sigs. A7v-A8r.

2. Thomas Dawson, *The Second Part of the Good Hus-wiues Iewell Where Is to Be Found Most Apt and Readiest Wayes to Distill Many Wholsome and Sweet Waters* (E. Allde for Edward White: London, 1597), pp. 39–40.

3. Anon., *A Closet for Ladies and Gentlewomen, or, The Art of Preseruing,*

Conseruing, and Candying (F. Kingston for Arthur Johnson: London, 1608), pp. 33–4; 38–9.

4. William Harrison, *The Description of England: The Classic Contemporary Account of Tudor Social Life,* ed. Georges Edelen (The Folger Shakespeare Library/Dover Publications: New York, 1994), p. 130.

5. David Dean, 'Elizabethan Government and Politics', in Robert Tittler and Norman Jones (eds.), *A Companion to Tudor Britain* (Blackwell: Oxford, 2004), p. 44.

6. Ibid., p. 44.

7. John Walter, *Crowds and Popular Politics in Early Modern England* (Manchester University Press: Manchester, 2010), p. 81.

8. Steve Rappaport, *Worlds Within Worlds: Structures of Life in Sixteenth-Century London* (Cambridge University Press: Cambridge, 1989), p. 137.

9. Ruth Goodman, *How to Be a Tudor: A Dawn-to-Dusk Guide to Tudor Life* (Liveright: New York, 2016), p. 149.

10. Sir Hugh Plat, *Sundrie New and Artificiall Remedies Against Famine* (Peter Short: London, 1596), sigs. B4v-C1r. This recipe is repeated in Sir Hugh Plat, *Delightes for Ladies to Adorne Their Persons, Tables, Closets and Distillatories* (Peter Short: London, 1602), sig. D7v.

11. Davidson, *The Oxford Companion to Food,* p. 147.

12. Nadiya Hussain, *Nadiya's Kitchen* (London, 2016), 'Parsnip and Orange Spiced Cake', *Happy Foodie,* https://thehappyfoodie.co.uk/recipes/parsnip-and-orange-spiced-cake.

13. Plat took inspiration here from the contemporary Latin book '*Anchora famis & sitis*', by Joachim Strupp von Gelnhausen in Plat, *Sundrie New and Artificiall Remedies Against Famine,* sig. D1v.

14. Ken Albala, *Food in Early Modern Europe* (Greenwood Press: Westport, CT, 2017), p. 62.

15. Althorp Papers, Add MS 75339, British Library, London (hereafter BL), ff. 18v–19r. 25 July 1639.

16. A. W., *A Booke of Cookrye. Very Necessary for All Such as Delight Therein* (Edward Allde: London, 1587), sig. C4v.

17. Elinor Fettiplace, *Elinor Fettiplace's Receipt Book,* ed. Hilary Spurling (Penguin: Harmondsworth, 1987), p. 206.

18. Goodman, *How to Be a Tudor,* p. 144.

19. Sue Shephard, *Pickled, Potted & Canned: How the Art and Science of Food Preserving Changed Civilisation* (Headline Books: London, 2000), p. 287.

20. According to John Aubrey, '*Brief Lives*', *Chiefly of Contemporaries, Set Down by John Aubrey, Between the Years 1669 & 1696,* ed. Andrew Clark (Clarendon Press: Oxford, 1898), pp. 75–6.

21. Helen Peavitt, *Refrigerator: The Story of Cool in the Kitchen* (Reaktion Books: London, 2017), p. 16.

22. Anon., *A Closet for Ladies and Gentlewomen*; Falsely attributed to Hannah Woolley, *The Accomplish'd Lady's Delight in Preserving, Physick, Beautifying, and Cookery* (B. Harris: London, 1675).

23. For some important examples see Sara Pennell, 'Perfecting Practice? Women, Manuscript Recipes and Knowledge in Early Modern England', in Victoria Burke and Jonathan Gibson (eds.), *Early Modern Women's Manuscript Writing* (Routledge: Aldershot, 2004), pp. 237–58; Catherine Field, '"Many Hands Hands": Writing the Self in Early Modern Women's Recipe Books', in Michelle M. Dowd and Julie A. Eckerle (eds.), *Genre and Women's Life Writing in Early Modern England* (Routledge: Aldershot, 2007), pp. 49–63.

24. Simon Werrett, 'Food, Thrift, and Experiment in Early Modern England', in Eleanor Barnett and Katrina Moseley (eds.), 'Special Issue: Food Waste and Sustainable Eating in Historical Perspective', *Global Food History*, 9 (2023), p. 226.

25. Fettiplace, *Elinor Fettiplace's Receipt Book*, p. 180.

26. As, for example, in John Murrell, *A Daily Exercise for Ladies and Gentlewomen* (T. Snodham for the widow Helme: London, 1617), sig. B2.

27. For example, Anon., *A Closet for Ladies and Gentlewomen*, pp. 8–9, 53–4; Murrell, *A Daily Exercise for Ladies and Gentlewomen*, sig. C7r.

28. Fettiplace, *Elinor Fettiplace's Receipt Book*, p. 66.

29. Thomas Dawson, *The Goodhuswifes Iewell Wherein is to Be Found Most Excellent and Rare Deuises for Conceits in Cookerie* (John Wolfe for Edward White: London, 1587), p. 16v.

30. Described as such by Samuel Pepys who said that they were 'now a great rarity since the war, none to be had'. Samuel Pepys, *The Diary of Samuel Pepys*, ed. Henry B. Wheatley (George Bell & Sons: London, 1893), 6 March 1665/66, http://www.gutenberg.org/files/4200/4200-h/4200-h.htm - link2H_4_0106.

31. Ibid., Monday 17 August 1668.

32. Fettiplace, *Elinor Fettiplace's Receipt Book*, p. 131.

33. Wendy Wall, 'Just a Spoonful of Sugar: Syrup and Domesticity in Early Modern England', *Modern Philology*, 104 (2006), p. 159; Victoria Avery, 'Sugar on the Table', in Victoria Avery and Melissa Calaresu (eds.), *Feast & Fast: The Art of Food in Europe 1500-1800* (Philip Wilson: London, 2019), p. 85.

34. Dawson, *The Second Part of the Good Hus-wiues Iewell*, p. 37; Plat, *Delightes for Ladies*, sig. C2r.

35. Wall, 'Just a Spoonful of Sugar: Syrup and Domesticity in Early Modern

England', p. 160.

36. Henry Buttes, *Dyets Dry Dinner Consisting of Eight Severall Courses* (Thomas Creede for William Wood: London, 1599), p. 229.

37. Ivan Day, 'Preparing and Preserving', in Avery and Calaresu (eds.), *Feast & Fast*, p. 117.

38. Henry VIII abolished the need to abstain from 'white meats' – meaning dairy products – during Lent. 'Dispensing with Lenten Fast from White Meats', 11 March 1538, in *Tudor Royal Proclamations, Volume 1: The Early Tudors (1485-1553)*, eds. Paul L. Hughes and James F. Larkin (Yale University Press: New Haven, CT, 1964), pp. 260–1. On food and fasting during the English Reformation, see Eleanor Barnett, 'Reforming Food and Eating in Protestant England, c. 1560 – c. 1640', *Historical Journal*, 63 (2020), pp. 507–27.

39. *Accompts of the Churchwardens of the Paryshe of St Christofer's in London 1572 to 1662*, ed. Edwin Freshfield (Rixon and Arnold: London, 1885), p. 76; 5 El. 1, c. 5 (1562).

40. Thomas Nash, *Nash's Lenten Stuff: Containing the Description and First Procreation and Increase of the Town of Great Yarmouth, in Norfolk*, ed. Charles Hindley (Reeves and Turner: London, 1871), pp. 42–3, 66. For a discussion of the debates surrounding foreign imports of salted fish see G. R. Elton, 'Piscatorial Politics in the Early Parliaments of Elizabeth I', in *Studies in Tudor and Stuart Politics and Government* (Cambridge, 2002), IV, pp. 109–30; R. C. L. Sgroi, 'Piscatorial Politics Revisited: The Language of Economic Debate and the Evolution of Fishing Policy in Elizabethan England', *Albion: A Quarterly Journal Concerned with British Studies*, 35 (2003), pp. 1–24.

41. Althorp Papers, Add MS 75339, BL.

42. Bridget Ann Henisch, *Fast and Feast: Food in Medieval Society* (Pennsylvania State University: University Park, PA, 1976), p. 40.

43. Thomas Tusser, *Fiue Hundreth Points of Good Husbandry Vnited to as Many of Good Huswiferie* (Richard Tottill: London, 1573), p. 12.

44. Ibid., p. 26.

45. Dawson, *The Goodhuswifes Iewell*, p. 26.

46. Anon., *A Closet for Ladies and Gentlewomen*, p. 65.

47. Plat, *Delightes for Ladies*, sig. C10r.

48. Dawson, *The Goodhuswifes Iewell*, p. 26v.

49. John Murrell, *A New Booke of Cookerie* (John Browne: London, 1615), pp. 2–3.

50. Plat, *Delightes for Ladies*, sig. C4r.

51. Pepys, *The Diary*, 22 August 1663.

52. As described by John Parkinson, *Paradisi in Sole Paradisus Terrestris*.

or, A Garden of All Sorts (Humphrey Lownes and Robert Young: London, 1629), p. 524.

53. Jennifer Stead, 'Necessities and Luxuries: Food Production from the Elizabethan to the Georgian Era', in C. Anne Wilson (ed.), *Waste Not, Want Not: Food Preservation from Early Times to the Present Day* (Edinburgh University Press: Edinburgh, 1991), p. 82.

54. Fettiplace, *Elinor Fettiplace's Receipt Book*, p. 144.

55. Stead, 'Necessities and Luxuries', p. 83.

56. Goodman, *How to Be a Tudor*, p. 120, 174.

57. As in Dawson, *The Goodhuswifes Iewell*, p. 19r.

58. Goodman, *How to Be a Tudor*, pp. 180–1.

59. Alison Sim, *The Tudor Housewife* (Sutton: Stroud, 2005), p. 71; Gervase Markham, *Countrey Contentments, or, The English Huswife* (John Beale for R. Jackson: London, 1623), p. 184.

60. An example of this theory comes from Philip Moore, *The Hope of Health Wherein is Conteined a Goodlie Regimente of Life* (John Kyngston: London, 1564), pp. xvii–xix.

61. John Ball, *Treatise of Faith Divided into Two Parts. The First Shewing the Nature, the Second, the Life of Faith* (William Stansby for Edward Brewster: London, 1631), p. 375.

62. Ibid., p. 377.

63. Richard Baxter, *A Christian Directory, or, A Summ of Practical Theologie and Cases of Conscience Directing Christians* (Robert White for Nevill Simmons: London, 1673), p. 146.

64. Thomas Leaver, *A Sermon Preached the Thyrd Sondaye in Lente Before the Kynges Maiestie, and His Honorable Counsell* (John Day: London, 1550); also see Samuel Rowlands, *A Sacred Memoire of the Miracles Wrought by Our Lord and Sauiour Iesus Christ* (Bernard Alsop: London, 1618), and Thomas Granger, *The Bread of Life, or Foode of the Regenerate. A Sermon Preached at Botterwike in Holland, Neere Boston, in Lincolnshire* (T. Snodham for Thomas Pavier: London, 1578). Matthew 14; Mark 6; Luke 9; John 6.

65. William Gouge, *Of Domesticall Duties: Eight Treatises* (William Bladen: London, 1622), p. 625.

66. Robert Cleaver, *A Godlie Form of Householde Governement for the Ordering of Priuate Families, According to the Direction of Gods Word* (Felix Kingston for Thomas Man: London, 1598), p. 384.

67. Ezekias Woodward, *A Childes Patrimony. Laid Out Upon the Good Nurture, or, Tilling Over the Whole Man* (I. Legat: London, 1640), pp. 56–9.

68. John Downame, *A Guide to Godlynesse, or, A Treatise of a Christian*

Life (F. Kingston for Philemon Stephens and Christopher Meredith: London, 1622), p. 295.

69. Luke 16:19–31.

70. Tusser, *Fiue Hundreth Points,* fol. 57v.

71. Markham, *Countrey Contentments,* p. 184.

72. Steve Hindle, 'Dearth, Fasting and Alms: The Campaign for General Hospitality in Late Elizabethan England', *Past and Present,* 172 (2001), pp. 44–5. See also *Acts of the Privy Council Volume 26, 1596–1597,* ed. John Roche Dasent (His Majesty's Stationery Office: London, 1902), *British History Online,* http://www.british-history.ac.uk/acts-privy-council, pp. 94–8; 380–2.

73. Lewis Bayly, *Practise of Pietie, Directing a Christian How to Walk That He May Please God* (John Hodgets: London, 1613), p. 648.

74. Terry Breverton, *The Tudor Kitchen: What the Tudors Ate & Drank* (Amberley Publishing: Stroud, 2015).

75. Felicity Heal, *Hospitality in Early Modern England* (Clarendon Press: Oxford, 1990), p. 34.

76. Paul S. Lloyd, *Food and Identity in England, 1540–1640: Eating to Impress* (Bloomsbury Academic: London, 2015), p. 33.

77. May, *The Accomplisht Cook,* sig. A6v.

78. Lloyd, *Food and Identity in England,* p. 32.

79. Mark Dawson, *Plenti and Grase: Food and Drink in a Sixteenth-Century Household* (Prospect Books: Totnes, 2009), p. 237.

80. Harrison, *The Description of England,* p. 127.

81. Markham, *Countrey Contentments,* p. 153.

82. Ibid., p. 39.

83. Frances E. Dolan, *Digging the Past: How and Why to Imagine Seventeenth-Century Agriculture* (University of Pennsylvania Press: Philadelphia, PA, 2020), p. 29; Ernesto Martin Uliarte et al., 'A Large-Scale Composting Experimentation Using Grape Marcs and Lees Residues from Mendoza Wine Alcohol Industry', Conference paper XXth Giesco International Meeting (2017).

84. Thomas Percy (ed.), *The Regulations and Establishment of the Household of Henry Algernon Percy, the Fifth Earl of Northumberland* (London, 1770), p. 57.

85. 'The Booke of the Household of Queene Elizabeth' (1601), in *A Collection of Ordinances and Regulations for the Government of the Royal Household, Made in Divers Reigns* (John Nichols for the Society of Antiquaries: London, 1790), pp. 281–98.

86. *Hertford County Records: Notes and Extracts from the Sessions Rolls 1581 to 1698,* I, ed. W. J. Hardy (C. E. Longmore: Hertford, 1905), pp. 35–6.

87. Leviticus 19: 9–10. And in Deuteronomy 24: 19–21. 'When thou beatest thine olive tree, thou shalt not go over the boughs again: it shall be for the stranger, for the fatherless, and for the widow. When thou gatherest the grapes of thy vineyard, thou shalt not glean it afterward: it shall be for the stranger, for the fatherless, and for the widow.' Biblical quotes from the Kings James Version (KJV).

88. Richard Bernard, *Ruths Recompence:, or, A Commentarie Vpon the Booke of Ruth* (Felix Kyngston: London, 1628), p. 134.

89. Peter King, 'Gleaners, Farmers and the Failure of Legal Sanctions in England, 1750–1850', *Past and Present*, 123 (1989), p. 142.

90. Stuart, *Waste*, p. 101.

91. *Hertford County Records*, p. 36.

92. See for example Francis Trigge, *To the Kings Most Excellent Maiestie. The Humble Petition of Two Sisters; The Church and Commonwealth* (George Bishop: London, 1604), sig. D2r.

93. Andy Wood, *The Memory of the People: Custom and Popular Senses of the Past in Early Modern England* (Cambridge University Press: Cambridge, 2013), p. 3.

94. Ibid., p. 4.

95. Bernard, *Ruths Recompence*, p. 136.

96. The Trussell Trust, 'Church Support', https://www.trusselltrust.org/get-involved/church-support/.

97. 'Ismaili Muslim Community Supports UK's Vulnerable With Joint 3.5 Tonne Food Donation', 19 November 2020, https://irp-cdn. multiscreensite.com/2fc87325/files/uploaded/FareShare%20Ismaili%20 CIVIC%20donation%20FINAL%2018%20November%202020.pdf.

98. Qu'ran 6: 141. English translation from Talal Itani in *Quran: English Translation. Easy to Read, in Modern English* (Ebook, 2014).

99. John Beale to Samuel Hartlib, 16 February 1657, The Hartlib Papers 62/24/1A-2B.

Chapter 2: Muckhills, Skins and Entrails

1. David L. Meeker and C. R. Hamilton, 'An Overview of the Rendering Industry', in David L. Meeker (ed.), *Essential Rendering: All About the Animal By-Products Industry* (National Renderers Association: Alexandria, VA, 2006), p. 2.

2. FABRA UK, 'Environmental Benefits of Rendering', http://www. fabrauk.co.uk/environmental-benefits-of-rendering More information can be found in FABRA factsheets, such as 'What Are

Animal By-Products', FABRA-FA-001, https://static1.squarespace.
com/static/5730558e5559862dca6ecc9a/t/605488d54ca0db68f6
9c04ca/1616152790316/FABRA-FS-001-What+are+Animal+By-
products-V1-+210308.pdf; 'Rendering – Process and Benefits', FABRA-
FS-002, https://
static1.squarespace.com/static/5730558e5559862dca6ecc9a/t/6447ce
d4dfcdde4ca67b9001/1682427605506/FABRA-FS-002-Rendering-
Process+and+Benefits-V4-230424%5B1%5D.pdf.; 'The Circular Economy
and Animal By-Products', FABRA-FS-004, https://static1.squarespace.
com/static/5730558e5559862dca6ecc9a/t/61c07ea9d0d4552c5cf8
7f13/1640005291410/FABRA-FS-004-The+Circular+Economy+and
+Animal+By-products-V2-211206.pdf. With many thanks to Nikki
Robertson from FABRA UK for kindly answering my questions.
3. See for example Dawson, *The Second Part of the Good Hus-wiues Iewell*,
p. 35; Peter Brears, 'Pots for Potting: English Pottery and its Role in
Food Preservation in the Post-Medieval Period', in Wilson (ed.), *Waste
Not, Want Not*, p. 32.
4. Receipt Book of Sarah Longe ca. 1610, Folger Shakespeare Library,
MS Add 444; Certain Profitable and Well Experienced Collections
for Making Conserve of Fruits [...] As Also of Surgery, Approved
Medicines ca. 1650, Folger Shakespeare Library, MS Add 259. See
Heather Wolfe, 'Early Modern Straws; or, Quills are Not Just for
Writing', *The Collation*, 2 October 2019, https://collation.folger.
edu/2019/10/quills-not-just-for-writing/.
5. Hannah Woolley, *The Ladies Directory in Choice Experiments &
Curiosities of Preserving in Jellies, and Candying Both Fruits & Flowers*
(T. M. for Peter Dring: London, 1662), p. 7. On p. 104 Woolley also
recommends using a feather to apply a wash onto sugar paste.
6. L. A. Clarkson, 'The Organization of the English Leather Industry in
the Late Sixteenth and Seventeenth Centuries, *The Economic History
Review,* 13 (1960), p. 246.
7. Jane Andrews, 'Industries in Kent, c. 1500–1640', in Michael Zell (ed.),
Early Modern Kent, 1540–1640 (The Boydell Press/Kent County Council:
Woodbridge, 2000), p. 134.
8. Lisa Yeomans, 'The Shifting Use of Animal Carcasses in Medieval and
Post-Medieval London', in Aleksander Pluskowski (ed.), *Breaking and
Shaping Beastly Bodies: Animals as Material Culture in the Middle Ages*
(Oxbow: Oxford, 2007) p. 101.
9. For example, 1548 2 & 3 Edw. 6, cap. 11; 1562 1 Eliz. I, cap. 8.
10. Clarkson, 'The Organization of the English Leather Industry in the Late
Sixteenth and Seventeenth Centuries', p. 246.

11. Richard B. Grantham, *A Treatise on Public Slaughter-Houses* (J. Weale: London, 1848), p. 9.
12. In the Tudor era most strings were imported from Italy and Germany. Later, 'catgut' strings, as they were known, were made in England until the early 1900s. Christopher Page, *The Guitar in Tudor England: A Social and Musical History* (Cambridge University Press: Cambridge, 2015), p. 75.
13. William Shakespeare, *Much Ado About Nothing*, 2, iii (1612).
14. Yeomans, 'The Shifting Use of Animal Carcasses in Medieval and Post-Medieval London', p. 112.
15. 'County: [Warwickshire]. Description of Courts: Manorial Courts', 29 April 1552.SC 2/207/82, National Archives, Kew, London.
16. City of Cambridge, *Whereas Divers Disordered People Inhabiting Amongst Us, Not Regarding the Good of this University, and Town of Cambridge* (Cambridge, 1635).
17. Peter Hounsell, *London's Rubbish: Two Centuries of Dirt, Dust and Disease in the Metropolis* (Amberley Publishing: Stroud, 2013), p. 10; Ernest L. Sabine, 'City Cleaning in Mediaeval London', *Speculum*, 12 (1937), pp. 22–3; Act of Common Council for Paving and Cleansing the Streets (1671).
18. Randle Holme, *The Academy of Armory, or, A Storehouse of Armory and Blazon* (Randle Holme: London, 1688), p. 334.
19. Sir Hugh Plat, *Diuerse New Sorts of Soyle*, in *The Iewell House of Art and Nature* (Peter Short: London, 1594), pp. 14–15, 33–4; John Evelyn, *A Philosophical Discourse of Earth Relating to the Culture and Improvement of It for Vegetation, and the Propagation of Plants* (John Martyn: London, 1676), pp. 97–9.
20. Gervase Markham, *Markhams Farewell to Husbandry or, The Inriching of All Sorts of Barren and Sterile Grounds in Our Kingdome* (M. Flesher for R. Jackson: London, 1625), pp. 57–62.
21. Evelyn, *A Philosophical Discourse of Earth*, p. 99.
22. David B. Goldstein, 'Manuring Eden: Biological Conversions in *Paradise Lost*', in Hillary Eklund (ed.), *Ground-Work: English Renaissance Literature and Soil Science* (Duguesne University Press: Pittsburgh, PA, 2017), pp. 183–93. For an example of the use of 'vegetative salt', see Adolphus Speed, *Adam out of Eden or, An Abstract of Divers Excellent Experiments Touching the Advancement of Husbandry* (Henry Brome: London, 1658), pp. 133–4.
23. Gervase Markham, *Cheap and Good Husbandry for the Well-Ordering of All Beasts and Fowls and for the General Cure of Their Diseases* (W. Wilson for George Sawbridge: London, 1664), p. 100.

24. John Stow, *A Survey of London. Reprinted from the Text of 1603*, ed. C. L. Kingsford (Clarendon: Oxford, 1908), pp. 175–86.

25. City of Cambridge, *Whereas Divers Disordered People Inhabiting Amongst Us.*

26. Act of Common Council for Paving and Cleansing the Streets (1671).

27. City of London, *The Lawes of the Markette* (John Cawoode: London, 1562), sig. A5v. These laws were repeatedly published throughout the seventeenth century; Thomas Almeroth-Williams, *City of Beasts: How Animals Shaped Georgian London* (Manchester University Press: Manchester, 2019), pp. 75–8.

28. Markham, *Cheap and Good Husbandry*, pp. 106–7.

29. Daniel Defoe, *A Tour Thro' the Whole Island of Great Britain* (G. Strahan: London, 1725), II, Letter 1, p. 48.

30. Robert Malcolmson and Stephanos Mastoris, *The English Pig: A History* (Hambledon Press: London, 1998), pp. 39–40.

31. Almeroth-Williams, *City of Beasts*, p. 85.

32. Malcolmson and Mastoris, *The English Pig*, p. 40.

33. Ian MacLachlan, 'A Bloody Offal Nuisance: The Persistence of Private Slaughter-houses in Nineteenth Century London', *Urban History*, 34 (2007), p. 236. The Butcher's Company received its Royal Charter of Incorporation in 1605 by James I.

34. Ernest L. Sabine, 'Butchering in Mediaeval London', *Speculum*, 8 (1933), p. 342.

35. Joseph Daw, *A Sketch of the Early History of the Worshipful Company of Butchers of London* (Worshipful Company of Butchers: London, 1869), p. 16.

36. Sabine, 'Butchering in Mediaeval London', p. 341.

37. Stow, *A Survey of London*, pp. 205–11.

38. 'Richard II: January 1393', in *Parliament Rolls of Medieval England*, ed. Chris Given-Wilson, et al. (Woodbridge, 2005), *British History Online*, http://www.british-history.ac.uk/no-series/parliament-rolls-medieval/january-1393; Daw, *A Sketch of the Early History of the Worshipful Company of Butchers of London*, p. 14. By an Act of Parliament in the reign of Charles I (1600–49) the butchers were bound to take all of the food waste to two 'Barrow-houses' on the Thames.

39. Peter Ackroyd, *Thames: Sacred River* (Chatto & Windus: London, 2007), p. 270.

40. Ibid., p. 270.

41. Anon., *At a Court of Sewers Held at the Guild Hall, London on Saterday the Fifth of February in the Year of Our Lord 1652* (Henry Hills for John Bellinger: London, 1653).

42. Stow, *A Survey of London*, p. 9.
43. François Jarrige and Thomas Le Roux, *The Contamination of the Earth: A History of Pollutions in the Industrial Age,* trans. Janice Egan and Michael Egan (The MIT Press: Cambridge, MA, 2020), p. 27.
44. Ackroyd, *Thames*, pp. 272–3.
45. City of London, *The Lawes of the Markette*, p. 3.
46. 'Lamentations of Old Father Thames', Harding BII (2049), V1290, Broadside Ballads Online, Bodleian Libraries, Oxford.
47. Andrew Boorde, *Hereafter Foloweth a Compendyous Regyment, or, A Dyetary of Helth* (Robert Wyer for John Gowghe: London, 1542), sigs. B4r-C1r.
48. Sabine, 'Butchering in Mediaeval London', p. 343.
49. Daw, *A Sketch of the Early History of the Worshipful Company of Butchers of London*, p. 20.
50. University of Cambridge, *Articles and Orders Agreed Vpon by the Right Worshipfull John Mansel Doctor of Divinite and Vicechancellor of the Vniversitie of Cambridge* (Cambridge, 1625). See also City of Cambridge, *Whereas Divers Disordered People Inhabiting Amongst Us.*
51. Thomas Cogan, *The Hauen of Health Cheifly Gathered for the Comfort of Students* (Anne Griffin for Roger Ball: London, 1584), p. 7.
52. Robert Burton, *The Anatomy of Melancholy, What It is with All The Kindes, Cavses, Symptomes, Prognostickes, and Severall Cyres of It* (John Lichfield and James Short for Henry Cripps: Oxford, 1621), pp. 108–9.
53. Thomas Tryon, *The Way to Health, Long Life and Happiness, or, A Discourse of Temperance* (H. C. for D. Newman: London, 1691), pp. 184–5.
54. Thomas Tryon, *Tryon's Letters Upon Several Occasions* (George Conyers and Elizabeth Harris: London, 1700), pp. 118, 121.
55. Edwin Chadwick, *Metropolitan Sewage Committee Proceedings. Parliamentary Papers* 1846, 10:651.
56. Peter Thorsheim, *Inventing Pollution: Coal, Smoke, and Culture in Britain Since 1800* (Ohio University Press: Athens, OH, 2006), p. 10.
57. City of London, *The Lawes of the Markette*, sig. A8v; Emily Cockayne, *Hubbub: Filth, Noise & Stench in England: 1600–1750* (Yale University Press: New Haven, CT, 2021), p. 95.
58. MacLachlan, 'A Bloody Offal Nuisance: The Persistence of Private Slaughter-houses in Nineteenth Century London', p. 234; Grantham, *A Treatise on Public Slaughter-Houses*, p. 17.
59. City of London, *The Lawes of the Markette*, pp. 37–8.
60. John Evelyn, *Fumifugium, or, The Inconvenience of the Aer and Smoak of London Dissipated* (W. Godbid: London, 1661), p. 13.

61. Cockayne, *Hubbub,* pp. 95–7.
62. Quoting the Butcher's Company's ordinance of 1607. Dorothy Davis, *A History of Shopping* (Routledge: London, 1966), pp. 85–6.
63. Anon., *The Experienced Market Man and Woman, or, Profitable Instructions, to All Masters and Mistrisses of Families, Servants and Others, to Know the Goodness of All Sorts of Provisions* (James Watson: Edinburgh, 1699), p. 6.
64. Adam Shewring, *The Plain-Dealing Poulterer* (C. Brome: London, 1687), pp. 5–6.
65. Cockayne, *Hubbub,* p. 96; Krish Seetah, 'The Middle Ages on the Block: Animals, Guilds and Meats in the Medieval Period', in Pluskowski (ed.), *Breaking and Shaping Beastly Bodies*, p. 21.
66. Cockayne, *Hubbub,* p. 101.
67. Sir Hugh Plat, *The Iewell House of Art and Nature* (Peter Short: London, 1594), pp. 22–3.
68. Markham, *Countrey Contentments,* p. 101.
69. Yeomans, 'The Shifting Use of Animal Carcasses in Medieval and Post-Medieval London', p. 100.
70. Grantham, *A Treatise on Public Slaughter-Houses,* p. 98.
71. Peter Atkins, 'Animal Wastes and Nuisances in Nineteenth-Century London', in Peter Atkins (ed.), *Animal Cities: Beastly Urban Histories,* (Routledge: London, 2016), p. 46.
72. 1847 10&11 Vict. c. 89.
73. MacLachlan, 'A Bloody Offal Nuisance: The Persistence of Private Slaughter-houses in Nineteenth Century London', p. 243.
74. Grantham, *A Treatise on Public Slaughter-Houses,* p. 83.
75. EU 2021/1372 of 17 August 2021, amending Annex IV to Regulation EC No 999/2001; Interview Tristram Stuart 10/12/2021.

Chapter 3: The World in a Tin Can

1. Mark Overton, *Agricultural Revolution in England: The Transformation of the Agrarian Economy 1500–1850* (Cambridge University Press: Cambridge, 1996), p. 125.
2. William Marshall, *The Rural Economy of the Midland Counties; Including the Management of Livestock in Leicestershire and its Environs,* I (G. Nicol: London, 1790), pp. 382, 391.
3. Jarrige and Le Roux, *The Contamination of the Earth,* p. 24.
4. Defoe, *A Tour Thro' the Whole Island of Great Britain,* II, Appendix, pp. 199–200.
5. Matthew White, 'The Industrial Revolution', *British Library,* 14 October

2009, https://www.bl.uk/georgian-britain/articles/the-industrial-revolution.

6. Joseph Boughey and Charles Hadfield, *British Canals: The Standard History* (The History Press: Cheltenham, 2022), p. 110.

7. Karen Foy, *Life in the Victorian Kitchen: Culinary Secrets and Servants' Stories* (Pen & Sword: Barnsley, 2014), p. 127.

8. Defoe, *A Tour Thro' the Whole Island of Great Britain*, I, Letter III, p. 60.

9. Lizzie Collingham, *The Hungry Empire: How Britain's Quest for Food Shaped the Modern World* (Random House: London, 2018), p. 119.

10. *Copies of Official Reports and Letters Relative to Donkin, Hall & Gamble's Preserved Provisions* (J. Powell: London, 1817), p. 10.

11. Shephard, *Pickled, Potted & Canned*, p. 231.

12. Nicholas Appert, *The Art of Preserving All Kinds of Animal and Vegetable Substances for Several Years* (Black, Parry and Kingsbury: London, 1811), pp. 1–2.

13. 'Copy of a Letter Written to General Caffarelli, Maritime Prefect at Brest, by the Council of Health, dated Brumaire, Year 12', Appert, *The Art of Preserving*, pp. xix–xx.

14. Hannah Woolley, *The Queen-like Closet, or, Rich Cabinet Stored with All Manner of Rare Receipts for Preserving, Candying & Cookery* (R. Lowndes: London, 1670), pp. 96, 151.

15. Shephard, *Pickled, Potted & Canned*, p. 228.

16. Appert, *The Art of Preserving*, pp. 42–3, 48.

17. Ibid., pp. 9–10.

18. Shephard, *Pickled, Potted & Canned*, p. 233.

19. John Gabriel Stedman, *Narrative, of a Five Years' Expedition; Against the Revolted Negroes of Surinam: in Guiana, on the Wild Coast of South America; From the Year 1772, to 1777* (J. Johnson: London, 1796), p. 121.

20. Shephard, *Pickled, Potted & Canned*, p. 231. In England, the first commercial manufacture of tinplate took place in 1699, whereas in France tinplate was not produced until 1720. Gordon L. Robertson, *Food Packaging: Principles and Practice*, 2nd edn. (CRC Press: Boca Raton, FL, 2006), p. 122.

21. William Parry, *Journal of a Voyage for the Discovery of a North-West Passage from the Atlantic to the Pacific* (John Murray: London, 1821), pp. 276–80.

22. Parry, *Journal of a Voyage*, p. 132.

23. Ibid., p. 212.

24. Ibid., p. 12.

25. *A Descriptive Catalogue of the Naval Manuscripts in the Pepysian Library at Magdalene College, Cambridge*, ed. J. R. Tanner (Navy

Records Society: London, 1903), p. 166.

26. Benjamin Franklin, *The Complete Works, in Philosophy, Politics, and Morals, of the Late Dr. Benjamin Franklin*, II (J. Johnson: London, 1806), p. 195.

27. Francis Grose, *A Classical Dictionary of the Vulgar Tongue* (S. Hooper: London, 1785), p. 104.

28. Plat, *Delightes for Ladies*, sig. F7r.

29. *London Daily Advertiser*, 1747.

30. Sloane MS 2244, British Library, London, f. 29.

31. Stead, 'Necessities and Luxuries', p. 84.

32. Anon., *The Lady's Companion: or, An Infallible Guide to the Fair Sex* (T. Read: London, 1743), pp. 116–17.

33. Stephen Bown, *Scurvy: How a Surgeon, a Mariner, and a Gentleman Solved the Greatest Medical Mystery of the Age of Sail* (Summersdale Publishers Ltd.: Chichester, 2003), p. 36.

34. Parry, *Journal of a Voyage*, pp. 178–9.

35. Appert, *The Art of Preserving*, p. 139.

36. As reported in *Grimsby Daily Telegraph*, 4 February 1939, and *Scunthorpe Evening Telegraph*, 4 February 1939.

37. William Parry, *Journal of a Third Voyage for the Discovery of a North-West Passage from the Atlantic to the Pacific; Performed in the Years 1824–25* (H. C. Carey and I. Lea: Philadelphia, PA, 1825), pp. 105–31.

38. Hansard, House of Commons, Debate: 'Preserved Meats (Navy)', V. 119, 12 February 1852.

39. Shephard, *Pickled, Potted & Canned*, p. 275.

40. Appert, *The Art of Preserving*, p. 161.

41. A. Wynter, *Our Social Bees, or, Pictures of Town & Country Life* (Robert Hardwicke: London, 1861), pp. 204–5.

42. Hansard, House of Commons, Debate: 'Preserved Meats (Navy)', V. 119, 12 February 1852.

43. Ibid.

44. Ibid.

45. Ibid.

46. Layinka Swinburne, 'Dancing with the Mermaids: Ship's Biscuit and Portable Soup', in Harlan Walker (ed.), *Food on the Move: Proceedings of the Oxford Symposium on Food and Cookery* (Prospect Books: Totnes, 1997), p. 312.

47. *Copies of Official Reports and Letters Relative to Donkin, Hall & Gamble's Preserved Provisions*, pp. 24–5, 42–3.

48. Beeton, *The Book of Household Management* (1861), p. 299.

49. Wynter, *Our Social Bees*, p. 199.

50. James Troubridge Critchell and Joseph Raymond, *A History of the Frozen Meat Trade: An Account of the Development and Present Day Methods of Preparation, Transport, and Marketing of Frozen and Chilled Meats* (Constable & Company Ltd.: London, 1912).

51. Dan Saladino, *Eating to Extinction: The World's Rarest Foods and Why We Need to Save Them* (Jonathan Cape: London, 2021), p. 149.

52. Peavitt, *Refrigerator*, pp. 42–3.

53. Beeton, *Mrs Beeton's Book of Household Management* (1907), p. 429.

54. Fyffes, 'Our Story', https://www.fyffes.com/our-story/our-history/.

55. B. H. Malyon, 'South African Shipping', *Journal of the Royal African Society*, 36 (1937), pp. 440–1.

56. As described by James Bostock, a Signaller in the 9th Bn DLI on the North African front. Second World War Experience Centre, 'Events in North Africa – June 1942', https://war-experience.org/events/events-in-north-africa-june-1942/.

57. H.M. Stationery Office, 'A.D. 1780 No. 1275: Preserving Vegetable Substances', *English Patents of Inventions, Specifications*, 3151–3280 (1857).

58. Malcolm C. Smith et al., 'Biomedical Results of Apollo', *NASA History Division*, https://history.nasa.gov/SP-368/s6ch1.htm.

Chapter 4: Mrs Beeton and the Rag-and-Bone Man

1. Henry Mayhew, *London Labour and the London Poor,* I (Griffin, Bohn, and Company: London, 1861), pp. 9–11, 28, 77.

2. Foy, *Life in the Victorian Kitchen*, p. 111.

3. MacLachlan, 'A Bloody Offal Nuisance: The Persistence of Private Slaughter-houses in Nineteenth Century London', p. 228; Gareth Stedman-Jones, *Outcast London: A Study in the Relationship Between Classes in Victorian Society* (Verso: London, 2013), p. 13.

4. Anne Murcott, *Introducing the Sociology of Food and Eating* (Bloomsbury Academic: London, 2019), p. 7.

5. Mayhew, *London Labour and the London Poor*, I, p. 129.

6. Almeroth-Williams, *City of Beasts*, p. 128.

7. Henry Mayhew, *London Labour and the London Poor,* III (Griffin, Bohn and Company: London, 1861), p. 1.

8. Kathryn Hughes, *The Short Life and Long Times of Mrs Beeton* (London, 2006), p. 195.

9. Mayhew, *London Labour and the London Poor*, I, pp. 4–5, 92.

10. Foy, *Life in the Victorian Kitchen*, p. 140; Phillis Browne, *The Girl's Own Cookery Book* (The Religious Tract Society: London, 1882), pp. 125–6.

11. Mayhew, *London Labour and the London Poor,* I, pp. 170–3.
12. Ibid., p. 117.
13. For example, 'Damaged and Ugly Fruit to Be Sold by Waitrose in a Bid to Cut Endemic Food Waste', *Daily Mail,* 2 June 2014, https://www.dailymail.co.uk/news/article-2645508/Damaged-ugly-fruit-sold-Waitrose-bid-cut-endemic-food-waste.html.
14. Mayhew, *London Labour and the London Poor,* I, pp. 104, 117–18.
15. Ibid., p. 179.
16. Charles Dickens, *Oliver Twist, or, The Parish Boy's Progress* (Project Gutenberg Ebook, 1996), Chapter L.
17. Henry Mayhew, 'A Visit to the Cholera Districts of Bermondsey', *The Morning Chronicle,* 24 September 1849.
18. Charles Dickens, *Bleak House* (Ticknor and Fields: Boston, 1867), p. 29.
19. Charles Dickens (Jr.), *Dictionary of London: An Unconventional Handbook* (Charles Dickens: London, 1879), p. 232.
20. Henry Mayhew, *London Labour and the London Poor,* II (Griffin, Bohn and Company: London, 1865), p. 124.
21. Overton, *Agricultural Revolution in England,* p. 111.
22. Mayhew, *London Labour and the London Poor,* II, pp. 146–9.
23. William Marshall Craig, *The Itinerant Traders of London* (Richard Phillips: London, 1804), n.p.n.
24. In Hounsell, *London's Rubbish,* pp. 19–20.
25. Mayhew, *London Labour and the London Poor,* II, pp. 186–97.
26. Ibid., pp. 154–8.
27. Beeton, *The Book of Household Management* (1861), p. 5.
28. Ibid., p. 2.
29. Alison Hulme, *A Brief History of Thrift* (Manchester University Press: Manchester, 2019), p. x.
30. Hughes, *The Short Life and Long Times of Mrs Beeton,* pp. 221–2.
31. Samuel Smiles, *Self-Help: With Illustrations of Character and Conduct* (John Murray: London, 1859), p. 230.
32. Beeton, *The Book of Household Management* (1861), p. 37.
33. Hughes, *The Short Life and Long Times of Mrs Beeton,* p. 149.
34. Beeton, *The Book of Household Management* (1861), p. 342.
35. Ibid., p. 37.
36. Ibid., p. 38.
37. Ibid., p. 816.
38. Ibid., p. 813.
39. Anon., *Murray's Modern Cookery Book. Modern Domestic Cookery* (John Murray: London, 1851), p. 12.
40. S. Beaty-Pownall, *Household Hints* (Horace Cox: London, 1904), p. 38.

41. Beeton, *Mrs Beeton's Book of Household Management* (1907), p. 430.
42. Hughes, *The Short Life and Long Times of Mrs Beeton*, p. 243.
43. John Burnett, *Plenty and Want: A Social History of Food in England from 1815 to the Present Day* (Routledge: Abingdon, 1989), p. 228.
44. Sale of Food and Drugs Act 1875; The Public Health (Preservatives etc. in Food) Regulations, 1925.
45. Beeton, *The Book of Household Management* (1861), pp. 372, 378.
46. Ibid., p. 24.
47. Ibid., p. 217.
48. For these later examples, see Eliza Acton, *Modern Cookery for Private Families* (Longmans: London, 1845), p. 162.
49. Beeton, *The Book of Household Management* (1861), pp. 226–7.
50. Ibid., p. 190.
51. Ibid., p. 522.
52. Ibid., p. 299.
53. Ibid., pp. 291–2.
54. Ibid., p. 292.
55. Ibid., p. 300.
56. Ibid., p. 37.
57. Ibid., p. 121.
58. Ibid., p. 749.
59. Ibid., pp. 816, 820.
60. Ibid., p. 913.
61. Ibid., p. 756.
62. Beeton, *Mrs Beeton's Book of Household Management* (1907), pp. 72–3.
63. Peavitt, *Refrigerator*, pp. 44–5.
64. Beeton, *Mrs Beeton's Book of Household Management* (1907), p. 89–90.
65. Beeton, *The Book of Household Management* (1861), pp. 88–9.
66. Beeton, *Mrs Beeton's Book of Household Management* (1907), pp. 783–4.
67. Ibid., pp. 779–80.
68. Michael French and Jim Phillips, 'Cheated Not Poisoned?: Food Regulation in the United Kingdom, 1875–1938' (Manchester University Press: Manchester, 2000), p. 97.
69. Beeton, *Mrs Beeton's Book of Household Management* (1907), p. 1296.
70. French and Phillips, *Cheated Not Poisoned*, p. 96.
71. The Public Health (Preservatives etc. in Food) Regulations, 1925.
72. For example, *Globe*, 30 November 1901, p. 8.
73. French and Phillips, *Cheated Not Poisoned*, pp. 36, 38.
74. Beeton, *The Book of Household Management* (1861), p. 178.
75. Beeton, *Mrs Beeton's Book of Household Management* (1907), p. 94, plate.
76. Licence, *What the Victorians Threw Away*, p. 13.

77. Ibid., p. 105.

78. William H. Maxwell, *The Removal and Disposal of Town Refuse* (The Sanitary Publishing Company Ltd.: London, 1898), p. 37.

79. W. F. Goodrich, *Economic Disposal of Towns' Refuse* (P. S. King & Son: London, 1901), p. 21.

80. Maxwell, *The Removal and Disposal of Town Refuse*, pp. 84–5; Thomas Codrington, *Report on the Destruction of Town Refuse* (Her Majesty's Stationery Office: London, 1888), pp. 3, 13, 48.

81. Hounsell, *London's Rubbish*, pp. 70–1.

Chapter 5: The Kitchen Front

1. Caroline Scott, *Holding the Home Front: The Women's Land Army in the First World War* (Pen & Sword: Barnsley, 2017), pp. 14, 39.

2. Alice Autumn Weinreb, *Modern Hungers: Food and Power in Twentieth-century Germany* (Oxford University Press: Oxford, 2017), p. 16.

3. Ibid., p. 27.

4. Quoted in Weinreb, *Modern Hungers*, p. 17.

5. Scott, *Holding the Home Front*, p. 124.

6. George V, 'By the King. A Proclamation', 2 May 1917.

7. National War Savings Committee, 'Mr. Slice O'Bread' (London, 1917).

8. Scott, *Holding the Home Front*, p. 230.

9. Lizzie Collingham, *The Taste of War: World War Two and the Battle for Food* (Allen Lane: London, 2011), pp. 114–15.

10. C. S. Lewis, *The Lion, the Witch and the Wardrobe* (1950, Macmillan: London, 1961 rpt.), p. 28.

11. Waste of Food Order, 1940, No. 1424, 5 August 1940. The order was revoked in 1953. The Waste of Food (Revocation) Order, 1953, No. 1589; Hansard, House of Commons Debate: 'Waste of Food Order (Prosecution)', V. 386, 27 January 1943.

12. Ian F. W. Beckett, *The Home Front 1914–1918: How Britain Survived the Great War* (Bloomsbury Information: London, 2006), p. 114.

13. Scott, *Holding the Home Front*, p. 194.

14. Imperial War Museum, 'Voices of the First World War: Life on the Home Front', https://www.iwm.org.uk/history/voices-of-the-first-world-war-life-on-the-home-front.

15. Dorothy Ann Haigh (Oral history), IWM Sound Archive, n. 734, (17/02/1976), reel 5.

16. Imperial War Museum, 'Voices of the First World War: Life on the Home Front'.

17. Scott, *Holding the Home Front,* pp. 195–6.
18. Burnett, *Plenty and Want,* p. 247.
19. Vere Hodgson, *Few Eggs and No Oranges: The Diaries of Vere Hodgson 1940–45* (Persephone Books: London, 2002), p. 231.
20. Lettice Miller (Oral History) IWM Sound Archive, n. 14241 (8/1994), reel 1.
21. Hodgson, *Few Eggs and No Oranges,* pp. 16, 82, 129, 228, 230, 285, 359, 471.
22. Ministry of Food, Leaflet Number 31, 'Cooking for One'.
23. British Pathé, War Archives, 'Before and After: Easter in Peacetime and Wartime' (1941), 23 Aug 2011, https://www.youtube.com/watch?v=_lmNyoxujac&ab_channel=WarArchives.
24. Ministry of Food, War Cookery Leaflet Number 4, 'Carrots'.
25. Ministry of Food, War Cookery Leaflet Number 4, 'Carrots' (July 1943).
26. Rebecca Earle, *Feeding the People: The Politics of the Potato* (Cambridge University Press: Cambridge, 2020), p. 44.
27. Hugh R. Rathbone, Letter to the *Liverpool Echo,* 12 January 1918, The National Archives, Kew, London, MAF 60/251b. See Chris Day, 'The Great War British Bake Off', The National Archives, 21 September 2015, https://blog.nationalarchives.gov.uk/great-war-british-bake/.
28. Ministry of Food, *Potato Pete's Recipe Book* (London, 1943).
29. Recording available online VintageBritishComedy, 'Elsie & Doris Waters – the Kitchen Front (20/12/1941), 4 June 2012, https://www.youtube.com/watch?v=snauq-Yglhs.
30. Joan Stokoe, 'Banana Sandwiches' by Northumberland County Libraries, WW2 People's War https://www.bbc.co.uk/history/ww2peopleswar/stories/03/a2734003.shtml.
31. Collingham, *The Taste of War,* p. 160.
32. Ministry of Food, War Cookery Leaflet Number 11, 'Dried Eggs'.
33. Collingham, *The Taste of War,* pp. 160–1.
34. Lesley Steinitz, 'Transforming Pig's Wash into Health Food: The Construction of Skimmed Milk Protein Powders', in Barnett and Moseley, 'Special Issue: Food Waste and Sustainable Eating in Historical Perspective', p. 293.
35. Kendra Smith-Howard, *Pure and Modern Milk: An Environmental History Since 1900* (Oxford University Press: Oxford, 2017), p. 67.
36. Ibid., pp. 72–5.
37. Hodgson, *Few Eggs and No Oranges,* pp. 231, 372.
38. Jack Tar Tuna Fish advert, *Truth,* London, 16 January 1918.
39. Collingham, *The Taste of War,* p. 159.
40. Ministry of Food, Leaflet Number 11, 'What's Left in the Larder',

(December 1946).

41. Ministry of Food, Leaflet Number 27, 'Potatoes' (October 1946).

42. Ministry of Food, *Potato Pete's Recipe Book.*

43. Ministry of Food, 'Making the Most of the Fat Ration' (April 1948).

44. British Pathé, 'Queen Visits Salvage Centres 1941' (3/4/1941), https://www.britishpathe.com/video/queen-visits-salvage-centres.

45. Peter Thorsheim, *Waste into Weapons: Recycling in Britain During the Second World War* (Cambridge University Press: Cambridge, 2015), pp. 88–9.

46. Mark Riley, 'From Salvage to Recycling – New Agendas or Same Old Rubbish?', *Area*, 40 (2008), p. 84.

47. British Pathé, 'Ministry of Information: Bones..Bones..Bones – Save Bones' (24/08/1944), https://www.britishpathe.com/video/bones-bones-bones-save-bones.

48. Weinreb, *Modern Hungers*, p. 18.

49. Saladino, *Eating to Extinction*, pp. 67–9.

50. Thorsheim, *Waste into Weapons*, p. 88.

51. Imperial War Museum, 'Pig Food: Women's Voluntary Service Collects Salvaged Kitchen Waste, East Barnet, Hertfordshire, England, 1943', https://www.iwm.org.uk/collections/item/object/205200094.

52. Thorsheim, *Waste into Weapons*, p. 73.

53. Julie Summers, *Jambusters: The Remarkable Story Which Has Inspired the ITV Drama Home Fires* (Simon & Schuster: London, 2013), p. 150.

54. Ibid., pp. 149–50.

55. Malcolmson and Mastoris, *The English Pig*, p. 49.

56. James Ross Cameron (introduced by), *Memory Lane: A Photographic Album of Daily Life in Britain 1930–1953* (J. M. Dent & Sons Ltd.: London, 1980), figure 194.

57. Joe Drury, 'War on Waste', *Guardian*, 16 May 2001, https://www.theguardian.com/society/2001/may/16/guardiansocietysupplement3.

58. Ministry of Food, Dig for Victory Leaflet Number 7, 'How to Make a Compost Heap' (1943).

59. Nigel Walford, 'The Extent and Impact of the 1940 and 1941 "Plough-up" Campaigns on Farming Across the South Downs, England', *Journal of Rural Studies*, 32 (2013), p. 38.

60. Collingham, *The Taste of War*, p. 146.

61. Ministry of Food, Dig for Victory Leaflet Number 16, 'Garden Pests and How to Deal with Them' (1941).

62. Jill Norman, *Eating for Victory: Healthy Home Front Cooking on War Rations* (Michael O'Mara Books Limited: London, 2014), Kindle edn, loc. 11 of 163.

63. Women's Land Army LAAS Handbook, *c.* 1917.
64. Annie Sarah Edwards (Oral history), IWM Sound Archive, n. 740 (4/3/1976), reels 1, 2, 7, 8.
65. Doris Robinson (Oral history), IWM Sound Archive, n. 12582 (1/6/1992), reel 1.
66. Scott, *Holding the Home Front,* pp. 154–5.
67. Margaret Mary Rumbold (Oral history), IWM Sound Archive, n. 8856 (31/05/1985), reels 1–4.
68. Margaret Mary Rumbold (Oral history), IWM Sound Archive, n. 8856 (31/05/1985), reels 3–4. On access to milk also see Iris Lilian Hobby (Oral history), IWM Sound Archive, n. 18274, reel 2.
69. Cherish Watton, 'Women's Land Army & Timber Corp', https://www.womenslandarmy.co.uk/first-world-war-womens-land-army/recruitment/.
70. National Army Museum, 'Women's Land Army Sorting Potatoes, 1944', https://collection.nam.ac.uk/detail.php?q=searchType%3Dsimple%26simpleText%3Dfood%26themeID%3D%26resultsDisplay%3Dlist%26page%3D8&pos=13&total=325&page=8&acc=1986-11-15-47.
71. Scott, *Holding the Home Front,* pp. 180–1.
72. Summers, *Jambusters,* p. 163.
73. Women's Institute, 'History', 'The WI: Inspiring Women', https://www.thewi.org.uk/about-us/history-of-the-wi.
74. Summers, *Jambusters,* p. 187.
75. Moira Meighn, 'Cooking for Victory', *The Land Girl,* Vol. 1, no. 5 (August 1940), p. 5.
76. Thorsheim, *Waste into Weapons,* p. 254.
77. Ibid., p. 107.
78. Collingham, *The Taste of War,* p. 189.
79. Ibid., pp. 217–32.
80. Ibid., pp. 686–7.

Chapter 6: McDonald's, Punks and Food Anarchists

1. With warm thanks to Keith McHenry for permission to reprint this recipe here.
2. Interview Ann Brinkworth and Sue Travers 4/11/2021.
3. Raymond G. Stokes, Roman Köster and Stephen C. Sambrook, *The Business of Waste: Great Britain and Germany, 1945 to the Present* (Cambridge University Press: Cambridge, 2013), p. 135; Mary Gwynn, *Back in Time for Dinner: From Spam to Sushi: How We've Changed the*

   Way We Eat (Random House: London, 2015), p. 52.
4. Interview Brinkworth and Travers 4/11/2021.
5. Gwynn, *Back in Time for Dinner*, p. 26.
6. Interview Brinkworth and Travers 4/11/2021.
7. Quoted in Gwynn, *Back in Time for Dinner*, p. 52.
8. Quoted in Peavitt, *Refrigerator*, pp. 131-2.
9. Ministry of Agriculture, Fisheries and Food, *Household Food Consumption and Expenditure: 1970 and 1971 with a Review of the Five Years 1966 to 1970: A Report of the National Food Survey Committee* (Her Majesty's Stationery Office: London, 1973), p. 2; Jon Cloke, 'Interrogating Waste: Vastogenic Regimes in the 21st Century', in Reynolds, Soma, Spring, and Lazell (eds.), *Routledge Handbook of Food Waste*, p. 71.
10. Mary Berry, *Popular Freezer Cookery* (Octopus Books: London, 1972), pp. 7–8.
11. Ibid.
12. Interview Brinkworth and Travers 4/11/2021.
13. Berry, *Popular Freezer Cookery*, pp. 50, 57, 59, 62, 66, 68–9, 72, 96.
14. David Evans, 'Binning, Gifting and Recovery: The Conditions of Disposal in Household Food Consumption', *Society and Space*, 30 (2012), p. 1132.
15. Stokes, Köster and Sambrook, *The Business of Waste*, p. 133.
16. Stuart, *Waste*, p. 77.
17. Collingham, *The Taste of War*, p. 705.
18. Stokes, Köster and Sambrook, *The Business of Waste*, p. 133.
19. Gwynn, *Back in Time for Dinner*, p. 48.
20. Cloke, 'Interrogating Waste: Vastogenic Regimes in the 21st Century', p. 70.
21. Ibid., pp. 71–2.
22. Tim Cooper, 'Challenging the "Refuse Revolution": War, Waste and the Rediscovery of Recycling, 1900–50', *Historical Research*, 81 (2008), p. 730.
23. Rachel Carson, *Silent Spring* (Penguin: London, 2000 edn.), esp. p. 22.
24. Ibid., pp. 24–5.
25. Ibid., esp. pp. 24–6, 31, 62–5.
26. Lord Shackleton, 'Introduction', in Carson, *Silent Spring*, pp. 11–12.
27. L. B. Powell, 'Spanner in the Soil: Why it is Time to Give up Industry for Good Husbandry', *The Ecologist*, 1:6 (December 1970), p. 24.
28. Michael Allaby, 'One Jump Ahead of Malthus', *The Ecologist*, 1:1 (July 1970), p. 26.
29. J. R. McNeill, *Something New Under the Sun: An Environmental History of the Twentieth-Century World* (WW Norton: New York, 2000), p. 227.

30. Allaby, 'One Jump Ahead of Malthus', p. 26.

31. 'Editorial', *The Ecologist*, 1:1 (July 1970), p. 4.

32. Allaby, 'One Jump Ahead of Malthus', pp. 26–7.

33. Friends of the Earth, 'Our History Campaigning Since 1971', https://friendsoftheearth.uk/who-we-are/our-history.

34. Hounsell, *London's Rubbish*, p. 154.

35. Andrew F. Smith, 'The Perfect Storm: A History of Food Waste', in Reynolds, Soma, Spring and Lazell (eds.), *Routledge Handbook of Food Waste*, p. 40.

36. Stokes, Köster and Sambrook, *The Business of Waste*, p. 106.

37. Jean Liefloff, 'A Lot of Rubbish', *The Ecologist*, 1:3 (September 1970), p. 37.

38. Patsy Hughes and Antony Seely, 'Making Waste Work', Landfill Research Paper 96/103, House of Commons (8 November 1996), p. 5.

39. Annechen Bahr Bugge, 'Fast Food', in Daniel Thomas Cook, J. Michael Ryan (eds.), *The Wiley Blackwell Encyclopedia of Consumption and Consumer Studies* (Wiley-Blackwell: Chichester, 2015), p. 289.

40. Owen's testimony comes from research initially carried out as part of an IHR-Museum of Youth Culture funded project. Eleanor Barnett, 'Fast Food: Eating and Youth Culture in Britain, 1970–2000', *Museum of Youth Culture*, 8 December 2020, https://museumofyouthculture.com/fast-food-youth-culture/.

41. London Greenpeace 'What's Wrong with McDonald's? Everything They Don't Want You to Know', 1986, https://www.mcspotlight.org/case/pretrial/factsheet.html.

42. Bob Langert, *The Battle to Do Good: Inside McDonald's Sustainability Journey* (Emerald Publishing: Bingley, 2019), p. 3.

43. Barnett, 'Fast Food: Eating and Youth Culture in Britain, 1970–2000'.

44. Spanner Films, 'McLibel: Two People Who Wouldn't Say Sorry' (2005).

45. Derek Oddy, *From Plain Fare to Fusion Food: British Diet from the 1890s to the 1990s* (The Boydell Press: Woodbridge, 2003), p. 231.

46. Langert, *The Battle to Do Good*, pp. 13–31, 111.

47. Dylan Clark, 'The Raw and the Rotten: Punk Cuisine', *Ethnology*, 43 (2004), p. 28.

48. McHenry, *Hungry for Peace*, p. 75.

49. Quoted in Keith McHenry, *The Anarchist Cookbook* (See Sharp Press: Tucson, AZ, 2015), p. 97.

50. C. T. Butler and Keith McHenry, *Food Not Bombs: How to Feed the Hungry and Build Community* (See Sharp Press: Tucson, AZ, 2000 edn.), https://www.foodnotbombs.net/bookindex.html, 'Why Food Not Bombs'; McHenry, *Hungry for Peace*, p. 58.

51. Butler & McHenry, *Food Not Bombs*, 'Logistics'.

52. Ibid.

53. Butler & McHenry, *Food Not Bombs*, 'Recipes'. With many thanks to Keith McHenry for permission to reprint this recipe.

54. Keith McHenry, Email exchange, September 2023.

55. McHenry, *The Anarchist Cookbook*, p. 125.

56. Ibid., pp. 141–2.

57. McHenry, *Hungry for Peace*, pp. 23, 41, 73.

58. Stuart, *Waste*, p. 13.

59. Interview Tristram Stuart 10/12/2021.

60. Stuart, *Waste*, pp. 17, 28.

61. Ibid., p. 22.

62. Amelia Gentleman, 'Prosecutors Drop Case Against Men Caught Taking Food from Iceland Bins', *The Guardian*, 29 January 2014, https://www.theguardian.com/uk-news/2014/jan/29/prosecutors-drop-case-men-food-iceland-bins.

63. Stuart, *Waste*, p. 21.

64. Feedback, *The Catering Toolkit: A Guide to Organizing Spectacular and Celebratory Public Events that Tackle Food Waste!*, https://feedbackglobal.org/wp-content/uploads/2016/12/F5K-Catering-toolkit.pdf, p. 55.

65. Interview Stuart 10/12/2021.

66. Feedback, 'Feeding the 5000', https://feedbackglobal.org/campaigns/feeding-the-5000/.

67. Feedback, 'About', *The Gleaning Network*, https://gleaning.feedbackglobal.org/about/.

68. Feedback, 'An Example Recipe to Feed 5000 People Per Guidelines', *The Catering Toolkit*, p. 55.

69. Feedback, 'Gleaning Network', https://feedbackglobal.org/campaigns/gleaning-network/.

70. WRAP, 'History of the Courtauld Commitment', https://wrap.org.uk/taking-action/food-drink/initiatives/courtauld-commitment/history-courtauld-commitment.

71. History Extra, 'From William Wordsworth to Extinction Rebellion: A History of Britain's Green Activists', 22 October 2019, https://www.historyextra.com/period/modern/william-wordsworth-extinction-rebellion-history-britains-green-activists-environment-campaigning-activism-environmentalism/.

72. Stuart, *Waste*, p. 149.

73. Eunomia for WRAP, *Dealing with Food Waste in the UK*, March 2007, pp. ii, 54.

74. As reported in the *Liverpool Echo*. 'Love Food Hate Waste – Broccoli

Stalk Soup', *Liverpool Echo*, 5 June 2009, https://www.liverpoolecho.co.uk/news/liverpool-news/love-food-hate-waste---3451592. The Love Food Hate Waste campaign continues to publish recipes on its website: WRAP, 'Love Food Hate Waste', https://www.lovefoodhatewaste.com/recipes.

75. With warm thanks to Robert Waller's daughter, Anne Baillie, for permitting me to print part of this poem here.

76. Robert Waller, 'Scabby Apples', *The Ecologist*, 1:4, (October 1971), p. 42.

77. Derek J. Oddy, 'From Roast Beef to Chicken Nuggets: How Technology Changed Meat Consumption in Britain in the Twentieth Century', in Derek Oddy and Alain Drouard (eds.), *The Food Industries of Europe in the Nineteenth and Twentieth Centuries* (Abingdon, 2016), p. 243.

78. John Burnett, *England Eats Out: A Social History of Eating Out in England from 1830 to the Present* (Routledge: Abingdon, 2014), p. 313.

79. WRAP report, quoted in Smith, 'The Perfect Storm: A History of Food Waste', p. 46.

Chapter 7: Food Waste in the Time of Coronavirus

1. Maria Nicola et al., 'The Socio-economic Implications of the Coronavirus Pandemic (COVID-19): A Review', *International Journal of Surgery*, 78 (June 2020), p. 190; S. Jack, 'What Are Shops Doing About Stockpiling?', BBC News, 22 March 2020, https://www.bbc.co.uk/news/business-51737030 (accessed March 2020).

2. 'Coronavirus: Nurse's Despair as Panic-Buyers Clear Shelves', *BBC News*, 19 March 2020, https://www.bbc.co.uk/news/av/uk-england-york-north-yorkshire-51966337.

3. 'Coronavirus: Ancient Mill Resumes Commercial Flour Production', *BBC News*, 22 April 2020, https://www.bbc.co.uk/news/uk-england-dorset-52369075.

4. 'Migrant Farmworkers Whose Harvests Feed Europe Are Blocked at Borders', *New York Times*, 27 March 2020, https://www.nytimes.com/2020/03/27/business/coronavirus-farm-labor-europe.html.

5. The Royal Family Channel, 'Prince Charles Urging People to "Pick for Britain"' (19 May 2020), https://www.youtube.com/watch?v=taHhUoxvBL8&ab_channel=TheRoyalFamilyChannel.

6. Sophie Gallagher, 'Pick for Britain: How to Get Involved with Campaign Backed by Royals', *The Independent*, 19 May 2020, https://www.independent.co.uk/life-style/pick-britain-how-sign-fruit-vegetable-harvest-a9521976.html.

7. Frankie Adkins, 'The Fruitless Saga of the UK's "Pick for Britain"

Scheme', *Al Jazeera*, 19 November 2020, https://www.aljazeera.com/features/2020/11/19/pick-for-britain-a-rather-fruitless.

8. Survey 'Food in the Age of Covid-19', June 2021. Carried out with Dr Katrina Moseley.

9. Bristol Bites Back Better, 'Fight Food Waste', https://www.goingforgoldbristol.co.uk/indi-bbbb-area/fight-food-waste/.

10. DEMOS for Food Standards Agency, 'Food in a Pandemic – From Renew Normal: The People's Commission on Life After Covid-19', March 2021, https://demos.co.uk/wp-content/uploads/2021/03/Food-in-a-Pandemic.pdf, pp. 47, 83.

11. Survey, 'How Has the Corona Pandemic Affected the Way You Eat?', March 2020. I received ninety-one responses from across the world on this basic survey that, from March 2020, allowed people to log their experiences of food in lockdown.

12. Survey 'Food in the Age of Covid-19', June 2021.

13. Boris Johnson, 'Prime Minister's Statement on Coronavirus (COVID-19): 23 March 2020', *GOV.UK,* https://www.gov.uk/government/speeches/pm-address-to-the-nation-on-coronavirus-23-march-2020.

14. On Shapps' comments, see Rowena Mason, 'No 10 Corrects "Shop Once a Week" Comment by Shapps', *The Guardian*, 31 March 2020, https://www.theguardian.com/world/2020/mar/31/no-10-slaps-down-shapps-over-shop-once-a-week-comment-coronavirus.

15. Felicity Lawrence, 'Millions to Need Food Aid in Days as Virus Exposes UK Supply', *The Guardian,* 27 March 2020, https://www.theguardian.com/world/2020/mar/27/millions-to-need-food-aid-in-days-as-virus-exposes-uk-supply.

16. Anon. Survey, 'How Has the Corona Pandemic Affected the Way You Eat?', 7 May 2020.

17. WRAP, 'Life Under Covid-19: Food Waste Attitudes and Behaviours in 2020', November 2020 https://wrap.org.uk/resources/report/life-under-covid-19-food-waste-attitudes-and-behaviours-2020.

18. Anon. Survey, 'How Has the Corona Pandemic Affected the Way You Eat?', 7 April 2020.

19. As reported in Survey 'Food in the Age of Covid-19', June 2021.

20. Lillian (F, 30–39). Survey, 'Food Waste and the Coronavirus Pandemic', November 2021. This was a more detailed and specific survey on food waste and the pandemic that I launched at the end of 2021.

21. DEMOS, 'Food in a Pandemic – From Renew Normal: The People's Commission on Life After Covid-19', p. 8.

22. Lourens Swanepoel. Survey, 'How Has the Corona Pandemic Affected

the Way You Eat?', 23 March 2020.

23. DEMOS, 'Food in a Pandemic – From Renew Normal: The People's Commission on Life After Covid-19', p. 51.

24. Jordon (M, 30–39), and Lillian (F, 30–39). Survey, 'Food Waste and the Coronavirus Pandemic', November 2021.

25. Anon. Survey, 'How Has the Corona Pandemic Affected the Way You Eat?', 23 March 2020.

26. WRAP, 'Life Under Covid-19: Food Waste Attitudes and Behaviours in 2020'.

27. Michelle Perrett, 'Coronavirus: Frozen Food Sales Boom by £285m', *Food Manufacture*, 8 July 2020, https://www.foodmanufacture.co.uk/Article/2020/07/08/Coronavirus-Frozen-food-sales-boom-during-lockdown-with-sales-up-285m.

28. WRAP, 'Life Under Covid-19: Food Waste Attitudes and Behaviours in 2020'.

29. Fabiola (F, 25–29). Survey, 'Food Waste and the Coronavirus Pandemic', November 2021.

30. Food Standards Agency, 'Covid-19 Consumer Tracker Waves 1–4, 2020, https://www.food.gov.uk/sites/default/files/media/document/covid-19-wave-1-4-report-final-mc.pdf, pp. 17-18.

31. Dorothy Barrick. Survey, 'How Has the Corona Pandemic Affected the Way You Eat?', 24 March 2020.

32. 'Bristol Going for Gold: Sustainable Food Places Submission', p. 89; Reasons given relate to the year 2019–2020. The Trussell Trust, 'What We Do: Our Vision is for a UK Without the Need for Food Banks', https://www.trusselltrust.org/what-we-do/. See also Effie Papargyropoulou, Kate Fearnyough, Charlotte Spring and Lucy Antal, 'The Future of Surplus Food Redistribution in the UK: Reimagining a "Win-Win" Scenario', *Food Policy*, 108 (2022), p. 2.

33. Social Metrics Commission, *Measuring Poverty* (July 2019), p. 5; Lang, *Feeding Britain*, pp. 175–6.

34. 'Bristol Going for Gold: Sustainable Food Places Submission', p. 89.

35. Food Standards Agency, 'Covid-19 Consumer Tracker Waves 1–4, p. 8.

36. DEMOS, 'Food in a Pandemic – From Renew Normal: The People's Commission on Life After Covid-19', p. 23.

37. As reported in Bright Harbour for Food Standards Agency, 'The Lived Experience of Food Insecurity Under Covid-19', July 2020, https://www.food.gov.uk/sites/default/files/media/document/fsa-food-insecurity-2020_-report-v5.pdf, p. 9.

38. YouGov poll commissioned by the Food Foundation, 25–6 March 2020, as reported in Felicity Lawrence, 'Families Borrowing to Buy Food a

Week into UK Lockdown', *The Guardian*, 28 March 2020, https://www. theguardian.com/society/2020/mar/28/families-borrowing-buy-food-week-of-lockdown.

39. Data from the Food Foundation, via DEMOS, 'Food in a Pandemic – From Renew Normal: The People's Commission on Life After Covid-19', p. 23.

40. Bright Harbour 'The Lived Experience of Food Insecurity Under Covid-19', p. 5.

41. Ibid., pp. 21–4.

42. Helen Davies, Retail Partners Senior Manager at FareShare, as described in 'How FareShare Help Brands Like Tesco Redistribute Surplus Food to Those in Need', *Table Talk Podcast*, 13 April 2021, https://audioboom.com/posts/7842371-how-fareshare-help-brands-like-tesco-redistribute-surplus-food-to-those-in-need.

43. Nicola Mackay, Community Food Programme Manager at Tesco as described in 'How FareShare Help Brands Like Tesco Redistribute Surplus Food to Those in Need', Table Talk Podcast, 13 April 2021, https://audioboom.com/posts/7842371-how-fareshare-help-brands-like-tesco-redistribute-surplus-food-to-those-in-need; Rebecca Smithers, 'Morrisons Gives Food Banks £10m During Coronavirus Outbreak', *The Guardian,* 30 March 2020, https://www.theguardian.com/world/2020/mar/30/morrisons-gives-food-banks-10m-during-coronavirus-outbreak.

44. Bristol Bites Back Better, 'Surplus Food as a Sustainable Solution', 19 April 2021, https://www.goingforgoldbristol.co.uk/surplus-food-as-a-sustainable-solution/.

45. Naomi (F, 25–29). Survey, 'Food Waste and the Coronavirus Pandemic', November 2021.

46. Bristol Going for Gold: Sustainable Food Places Submission', p. 52.

47. Interview Phoebe Ruxton 30/11/2021.

48. 'Bristol Going for Gold: Sustainable Food Places Submission', p. 52.

49. FareShare South West, 'Impact Report 2021', March 2021, https://issuu.com/faresharesouthwest/docs/fareshare_impact_report_2021_digital, p. 5.

50. Interview Josh Eggleton 9/12/2021.

51. 'Bristol Going for Gold: Sustainable Food Places Submission', p. 92.

52. Bristol Bites Back Better, 'During the First Lockdown, Food Club Was a Lifesaver as I Was Unable to Go Out to the Shops', 13 March 2021, https://www.goingforgoldbristol.co.uk/during-the-first-lockdown-food-club-was-a-lifesaver-as-i-was-unable-to-go-out-to-the-shops/.

53. Papargyropoulou et al., 'The Future of Surplus Food Redistribution in

the UK: Reimagining a "Win-Win" Scenario', p. 2.

54. Jordon Lazell (M, 30–39). Survey, 'Food Waste and the Coronavirus Pandemic', November 2021.

55. Elizabeth Elkin, Mai Ngoc Chau and Agnieszka de Sousa, 'Your Food Prices Are at Risk as the World Runs Short of Workers', *Bloomberg*, 2 September 2021, https://www.bloomberg.com/news/features/2021-09-02/food-prices-driven-up-by-global-worker-shortage-brexit.

56. 'How Serious is the Shortage of Lorry Drivers', *BBC News*, 15 October 2021, https://www.bbc.co.uk/news/57810729.

57. 'Dairy Farmers Forced to Pour Tens of Thousands of Litres of Milk Away Due to HGV Driver Shortage and Rising Costs and Labour Shortages', *Sky News*, 6 October 2021, https://news.sky.com/story/dairy-farmers-forced-to-pour-tens-of-thousands-of-litres-of-milk-away-due-to-hgv-driver-shortage-and-rising-costs-and-labour-shortages-12427818.

58. Aimi Redfern, '"100,000 Litres of Milk Wasted" as Lorry Driver Shortage Hits Dairy Farmers', *StokeonTrentLive*, 10 September 2021, https://www.stokesentinel.co.uk/news/stoke-on-trent-news/100000-litres-milk-wasted-lorry-5893962.

59. Claire Marshall, 'Abattoir Labour Shortage Sees Yorkshire Farmer Kill Piglets', *BBC News*, 1 October 2021, https://www.bbc.co.uk/news/science-environment-58749841.

60. Alistair Driver, 'On-Farm Pig Cull Will Get "Worse and Worse" Unless Summit Delivers Results, Zoe Tells Today Programme', *National Pig Association*, 10 February 2022, http://www.npa-uk.org.uk/On-farm_pig_cull_will_get_worse_and_worse_unless_summit_delivers_results_Zoe_tells_Today_programme.html; Jonathan Riley, 'Crisis in Pig Sector Deepens as Welfare Cull Tops 30,000', *Farmers Weekly*, 16 December 2021, https://www.fwi.co.uk/livestock/crisis-in-pig-sector-deepens-as-welfare-cull-tops-30000.

61. Charlie Clutterbuck, *Bittersweet Brexit: The Future of Food, Farming, Land and Labour* (Pluto Press: London, 2017), p. 6.

62. Lang, *Feeding Britain*, pp. 188–9.

63. Clutterbuck, *Bittersweet Brexit*, pp. 84–5.

64. Ibid., p. 3.

65. Stuart, *Waste*, p. 133.

66. Rachel Schraer, 'EU Fishing Rules: Did the UK Throw Away a Million Tonnes of Fish?', *BBC News*, 11 February 2020, https://www.bbc.co.uk/news/uk-51415240.

67. UK Government, 'Statutory Guidance: Landing Obligation General

Requirements', *GOV.UK,* 21 July 2021, https://www.gov.uk/government/publications/technical-conservation-and-landing-obligation-rules-and-regulations-2021/landing-obligation-general-requirements#quota-limits; UK Government, 'Statutory Guidance: Landing Obligation General Requirements', GOV.UK, Updated April 2023, https://assets.publishing.service.gov.uk/media/643816a9773a8a0013ab2c0c/Landing_obligation_general_requirements_v2023.pdf..

68. 'Brexit: Why Is There a Row Over Fishing Rights?', *BBC News,* 23 December 2021, https://www.bbc.co.uk/news/46401558.

69. UK government statistics quoted in Jonty Bloom, 'How are Food Supply Networks Coping with Coronavirus?', *BBC News,* 26 March 2020, https://www.bbc.co.uk/news/business-52020648.

70. Adam Forrest, 'Brexit Checks on EU Food Imports Scrapped, Announces Jacob Rees-Mogg', *The Independent,* 28 April 2022, https://www.independent.co.uk/news/uk/politics/brexit-eu-import-checks-rees-mogg-b2067421.html.

71. Video tweeted by @10DowningStreet, 17 June 2020, 10.36 a.m.

72. Lang, *Feeding Britain,* pp. 41, 84–5, 233.

73. Clutterbuck, *Bittersweet Brexit,* pp. 55–68.

74. Ibid., p. 16.

75. Ibid., p. 34.

76. Sarah Coe and Jonathan Finlay, 'Agricultural Act 2020', *House of Commons Library,* Briefing Paper, 3 December 2020, https://researchbriefings.files.parliament.uk/documents/CBP-8702/CBP-8702.pdf

77. UK government statistics quoted in Bloom, 'How are Food Supply Networks Coping with Coronavirus?'.

78. DEMOS, 'Food in a Pandemic – From Renew Normal: The People's Commission on Life After Covid-19', p. 91.

79. Ibid., p. 5.

80. Food Ethics Council, 'Ethics in Our Food Response to Covid-19' (March 2020), https://www.foodethicscouncil.org/ethics-in-our-food-response-to-covid-19/.

81. 'Bristol Going for Gold: Sustainable Food Places Submission', p. 1.

82. See Bristol Bites Back Better, 'Fight Food Waste'.

83. 'Bristol Going for Gold: Sustainable Food Places Submission', p. 16.

84. Defra Press Office, 'Household Food Waste to Be Collected Separately by 2023 and 50,000 City Trees to Be Planted in Urban Tree Challenge Fund', 10 February 2020, https://deframedia.blog.gov.uk/2020/02/10/household-food-waste-to-be-collected-separately-by-2023-and-50000-city-trees-to-be-planted-in-urban-tree-challenge-fund/.

85. Ben Thomson, 'Changes to Food Waste Collection and Recycling: What Actions do Councils Need to Take?', *LocalGov*, 17 May 2022, https://www.localgov.co.uk/Changes-to-food-waste-collection-and-recycling-What-actions-do-councils-need-to-take-/54208; UK Government, 'Environment Act 2021', 57 'Managing Waste', esp. 45A(7).
86. Elena Ares and Agnieszka Suchenia, 'Allocations to UK-EU Fisheries Following the UK's Departure from the EU', *House of Commons Library*, 29 November 2021, https://researchbriefings.files.parliament.uk/documents/CDP-2021-0202/CDP-2021-0202.pdf.
87. Esyllt Carr, 'Supermarkets Set Limits on Sale of Cooking Oil', *BBC News*, 22 April 2022, https://www.bbc.co.uk/news/business-61R193141.
88. WRAP, 'Life Under Covid-19: Food Waste Attitudes and Behaviours in 2020'; WRAP, 'Returning to Normality After Covid-19: Food Waste Attitudes and Behaviours in 2021', August 2021, https://wrap.org.uk/sites/default/files/2021-08/food-trends-report-august-2021.pdf; WRAP, 'UK Household Food Waste Tracking Survey Winter 2021: Behaviours, Attitudes, and Awareness', February 2022, https://wrap.org.uk/sites/default/files/2022-03/WRAP-UK-household-food-waste-Winter-2021-Behaviours-attitudes-and-awareness.pdf.

Epilogue: A Food Waste Free Future?

1. IPCC, 'Summary for Policymakers: Special Report on Climate Change and Land', January 2020, https://www.ipcc.ch/srccl/chapter/summary-for-policymakers/.
2. WRAP, 'Life Under Covid-19: Food Waste Attitudes and Behaviours in 2020'.
3. IPCC, 'Summary for Policymakers: Special Report on Climate Change and Land'.
4. As of 2022. Action Against Hunger, 'World Hunger Facts', https://www.actionagainsthunger.org.uk/why-hunger/world-hunger-facts#:~:text=Globally%2C%20one%20in%20nine%20people%20are%20hungry%20or%20undernourished.&text=2.37%20billion%20people%20did%20not,and%20nutritious%20food%20in%202020.
5. Gunders, *Waste Free Kitchen Handbook*, p. 19.
6. Interview Stuart 10/12/2021.
7. Felicitas Schneider and Mattias Eriksson, 'Food Waste (and Loss) at the Retail Level', in Reynolds et al. (eds.), *Routledge Handbook of Food Waste*, p. 122.
8. Marcus Glover, 'Courtauld: A Model for the World to Follow in the

Fight Against Food Waste', *WRAP*, 8 October 2020, https://wrap.org.uk/blog/2020/10/courtauld-model-world-follow-fight-against-food-waste.

9. Jeswani, Figueroa-Torres and Azapagic, 'The Extent of Food Waste Generation in the UK and its Environmental Impacts', p. 545; Interview Stuart 10/12/2021.

10. Estelle Uba, 'Ministers Criticised for Scrapping New Food Waste Laws for England', *The Guardian*, 17 August 2023, https://www.theguardian.com/environment/2023/aug/17/ministers-criticised-for-scrapping-new-food-waste-laws-for-england; Ian Quinn, 'Mandatory Food Waste Reporting Regulation Scrapped', *The Grocer*, 28 July 2023, https://www.thegrocer.co.uk/food-waste/mandatory-food-waste-reporting-regulation-scrapped/681656.article.

11. Anne (F, 40s), Survey, 'Food Waste and the Coronavirus Pandemic', November 2021.

12. UK Government, 'Environment Act 2021', 57 'Managing Waste', 45A(4, 7); Defra Press Office, 'Household Food Waste to Be Collected Separately by 2023 and 50,000 City Trees to Be Planted in Urban Tree Challenge Fund', 10 February 2020, https://deframedia.blog.gov.uk/2020/02/10/household-food-waste-to-be-collected-separately-by-2023-and-50000-city-trees-to-be-planted-in-urban-tree-challenge-fund/; UK Government, 'Press Release: New Plans Unveiled to Boost Recycling', 7 May 2021, https://www.gov.uk/government/news/new-plans-unveiled-to-boost-recycling.

13. 'Bristol Going for Gold: Sustainable Food Places Submission', p. 47.

14. More information can be found here: WRAP, 'Love Food Hate Waste Campaigns', https://wrap.org.uk/taking-action/citizen-behaviour-change/love-food-hate-waste/key-campaigns (accessed June 2022).

15. Smith, 'The Perfect Storm: A History of Food Waste', pp. 42–3.

16. UK Government, 'Environment Act 2021', 57 'Managing Waste', 45AZB; Lisa Gilligan and Kirstin Roberts, 'Managing Separate Food Waste Segregation for All Businesses Set to Come into Force from 2023', *FREETHS*, 17 April 2023, https://www.freeths.co.uk/2023/04/17/mandatory-separate-food-waste-segregation-for-all-businesses-set-to-come-into-force-from-2023/.

17. According to the CEO Lindsay Boswell, FareShare, 'Budget: FareSare "Deeply Disappointed" No Funding Announced for Food Waste Scheme', 27 October 2021, https://fareshare.org.uk/news-media/press-releases/budget-fareshare-deeply-disappointed/.

18. ReFED 'Investment Tracker', https://refed.org/investment-tracker/ (accessed April 2022).

19. Elina Närvänen et al., 'Introduction: A Framework for Managing Food Waste', in Närvänen et al. (eds.), *Food Waste Management*, p. 4.
20. Apeel, 'Comparison to Current Methods for Preservation of Raw Fruits and Vegetables', https://apeelincanada.wordpress.com/2018/11/29/comparison-to-refrigeration-and-controlled-atmosphere-storage/.
21. Interview Moody Soliman 16/12/21.
22. For more information, see Neurolabs, 'Technology', https://www.neurolabs.ai/synthetic-data.
23. Wasteless, https://www.wasteless.com/.
24. Winnow, 'IKEA and Winnow Are Building the Kitchen of the Future', https://info.winnowsolutions.com/ikea-and-winnow-are-building-the-kitchen-of-the-future.
25. Interview David Jackson (Marketing Director at Winnow Solutions) 15/06/22.
26. Olio, 'Our Impact', https://olioex.com/about/our-impact/.
27. Zero Waste Europe, 'The Story of Too Good To Go', (2020), https://zerowasteeurope.eu/wp-content/uploads/2020/01/zero_waste_europe_CS7_CP_TooGoodToGo_en.pdf .
28. Interview Solveiga Pakštaitė 21/01/22.
29. Mic Wright, 'Has "Best Before" Reached its Sell-By Date?', *The Guardian*, 16 January 2009, https://www.theguardian.com/lifeandstyle/wordofmouth/2009/jun/16/food-waste-best-before-dates.
30. See, for example, Jasper Jolly, '"Use the Sniff Test": Morrisons to Scrap "Use By" Dates from Milk Packaging', *The Guardian*, 9 January 2022, https://www.theguardian.com/business/2022/jan/09/use-the-sniff-test-morrisons-to-scrap-use-by-dates-from-milk-packaging.
31. Adam Durbin, 'Co-op Supermarket Scraps Yoghurt Use-By Dates in Bid to Cut Food Waste', *BBC News*, 22 April 2022, https://www.bbc.co.uk/news/uk-61184855.
32. Smith, 'The Perfect Storm: A History of Food Waste', p. 42.
33. Martin Bowman, Karen Luyckx and Christina O'Sullivan, 'Keeping Unavoidable Food Waste in the Food Chain as Animal Feed', in Reynolds et al. (eds.), *Routledge Handbook of Food Waste*, p. 366.
34. Trevor M. Fowles and Christian Nansen, 'Insect-Based Bioconversion: Value from Food Waste', in Närvänen et al. (eds.), *Food Waste Management*, pp. 321–46.
35. Interview Stuart 16/12/21.
36. FAO, 'Beauty (and Taste!) Are on the Inside'.
37. Stephen D. Porter et al., 'Avoidable Food Losses and Associated Production-Phase Greenhouse Gas Emissions Arising from Application of Cosmetic Standards to Fresh Fruit and Vegetables in Europe and the

UK', *Journal of Cleaner Production,* 201 (2018), pp. 869–78.

38. Rubies in the Rubble, 'Our Story', https://rubiesintherubble.com/pages/our-story.

39. Andrea D. Steffen, 'New Processes Turn Dairy Waste into Bioplastics and Fertilizers', *Intelligent Living,* 23 July 2020, https://www.intelligentliving.co/process-dairy-waste/.

40. Larissa Zimberoff, *Technically Food: Inside Silicon Valley's Mission to Change What We Eat* (Abrams Press: New York, 2021), p. 101. I am grateful to Larissa for kindly sending me a copy of her book.

41. Cajsa Carlson, 'Valdís Steinarsdóttir Turns Animal Skin and Bones into Food Packaging and Vases', *Dezeen,* 27 January 2021, https://www.dezeen.com/2021/01/27/valdis-steinarsdottir-food-packaging-vessels-animal-skin-bones/.

42. Berry, *Popular Freezer Cookery,* p. 25.

43. Gunders, *Waste Free Kitchen Handbook,* p. 105.

44. Stuart, *Waste,* p. 147.

45. In 2019/20. Brigid Francis-Devine, 'Income Inequality in the UK', House of Commons Library, 30 November 2021, p. 19.

46. John-Paul Ford Rojas, 'Tory MP Criticised for Saying Food Bank Users Just Need to Learn How to Cook', 11 May 2022, https://news.sky.com/story/tory-mp-criticised-for-saying-food-bank-users-just-need-to-learn-how-to-cook-12610728.

47. Food and Agricultural Organization of the United States, 'Global Food Losses and Food Waste', 2011, https://www.fao.org/3/i2697e/i2697e.pdf, p. 5.

48. Silo, 'Becoming Zero Waste', https://silolondon.com/story/.

49. Zimberoff, *Technically Food,* pp. 88–9.

50. See Bristol Early Years, 'The Children's Kitchen', https://www.bristolearlyyears.org.uk/health/the-childrens-kitchen/.

51. Woodward, *A Childes Patrimony,* p. 56.

Timeline

1. Peter Brears, 'Pots for Potting: English Pottery and its Role in Food Preservation in the Post-Medieval Period', in Wilson (ed.), *Waste Not, Want Not,* p. 56.

INDEX

Page numbers in *italic* refer to Figures and Plates; fn after a page number refers to a footnote.